Thermal Uses and Properties
of Carbohydrates and Lignins

Symposium on Thermal Uses and Properties
of Carbohydrates and Lignins
172nd National Meeting of the American Chemical
Society, San Francisco, September, 1976

Thermal Uses and Properties of Carbohydrates and Lignins

Edited by

Fred Shafizadeh

Wood Chemistry Laboratory
University of Montana
Missoula, Montana

Kyosti V. Sarkanen

College of Forest Resources
University of Washington
Seattle, Washington

David A. Tillman

Materials Associates, Inc.
Washington, D.C.

ACADEMIC PRESS

New York   San Francisco   London   1976
A Subsidiary of Harcourt Brace Jovanovich, Publishers

ACADEMIC PRESS, INC.
111 Fifth Avenue, New York, New York 10003

United Kingdom Edition published by
ACADEMIC PRESS, INC. (LONDON) LTD.
24/28 Oval Road, London NW1

LIBRARY OF CONGRESS CATALOG CARD NUMBER: 76-55057

ISBN 0–12–637750–2

PRINTED IN THE UNITED STATES OF AMERICA

Contents

List of Contributors

HERBERT R. APPELL, *Pittsburgh Energy Research Center, U.S. Energy Research and Development Administration, 4800 Forbes Avenue, Pittsburgh, Pennsylvania 15213*

DAVID L. BRINK, *Forest Products Laboratory, University of California, 1301 So. 46th Street, Richmond, California 94804*

A. BROIDO, *Pacific Southwest Forest and Range Experiment Station, Forest Service, U.S. Department of Agriculture, Berkeley, California 94701*

H.R. BURTON, *Department of Agronomy, University of Kentucky, Lexington, Kentucky 40506*

K. E. CABRADILLA, *Department of Clothing, Textiles, and Interior Design, Kansas State University, Manhattan, Kansas 66506*

JAMES A. CHARLEY, *Forest Products Laboratory, University of California, 1301 So. 46th Street, Richmond, California 94804*

PAUL H. CHEN, *Philip Morris Research Center, P. O. Box 26583, Richmond, Virginia 23261*

WILLIAM F. DeGROOT, *Wood Chemistry Laboratory, Department of Chemistry, University of Montana, Missoula Montana 59801*

I. N. EINHORN, *Flammability Research Center, Department of Chemistry, University of Utah, Salt Lake City, Utah 84112*

GEORGE W. FALTICO, *Forest Products Laboratory, University of California, 1301 So. 46th Street, Richmond, California 94804*

P. FANG, *Forest Products Utilization Laboratory, Mississippi State University, P. O. Drawer FP, Mississippi State, Mississippi 39762*

M. D. FRAZER, *InterTechnology Corporation, 100 Main Street, Warrenton, Virginia 22186*

J. H. FUTRELL, *Flammability Research Center, Department of Chemistry, University of Utah, Salt Lake City, Utah 84112*

D. W. GOHEEN, *Crown Zellerbach Corporation, Camas, Washington 98607*

J. F. HENRY, *InterTechnology Corporation, 100 Main Street, Warrenton, Virginia 22186*

F. D. HILEMAN, *Flammability Research Center, Department of Chemistry, University of Utah, Salt Lake City, Utah 84112*

WILLIAM R. JOHNSON, *Philip Morris Research Center, P. O. Box 26583, Richmond, Virginia 23261*

JOHN C. KANG, Philip Morris Research Center, P. O. Box 26583, Richmond, Virginia 23261

J. A. KNIGHT, Engineering Experiment Station, Georgia Institute of Technology, Atlanta, Georgia 30332

JOHN M. LEAVENS, The Ralph M. Parsons Company, 100 West Walnut Avenue, Pasadena, California 91124

G. D. McGINNIS, Forest Products Utilization Laboratory, Mississippi State University, P. O. Drawer FP, Mississippi State, Mississippi 39762

J. V. ORLE, Crown Zellerbach Corporation, Camas, Washington 98607

PETER PANTAGES, Pittsburgh Energy Research Center, U.S. Energy Research and Development Administration, 4800 Forbes Avenue, Pittsburgh, Pennsylvania 15213

R. C. ROTHERMEL, Intermountain Forest and Range Experiment Station, Northern Forest Fire Laboratory, P. O. Drawer G., Missoula, Montana 59801

K. V. SARKANEN, College of Forest Resources, University of Washington, Seattle, Washington 98195

FRED SHAFIZADEH, Wood Chemistry Laboratory, Department of Chemistry, University of Montana, Missoula, Montana 59801

CORNELIUS STEELINK, Department of Chemistry, University of Arizona, Tucson, Arizona 85721

JEROME F. THOMAS, Forest Products Laboratory, University of California, 1301 So. 46th Street, Richmond, California 94804

DAVID A. TILLMAN, Materials Associates, Inc., 600 New Hampshire Ave., NW, Washington, D.C. 20037

C. W. VAIL, InterTechnology Corporation, 100 Main Street, Warrenton, Virginia 22186

ALAN WILKINSON, Department of Chemistry, University of Arizona, Tucson, Arizona 85721

E. MILTON WILSON, The Ralph M. Parsons Company, 100 West Walnut Avenue, Pasadena, California 91124

R. P. WITHER, Crown Zellerbach Corporation, Camas, Washington 98607

L. H. WOJCIK, Flammability Research Center, Department of Chemistry, University of Utah, Salt Lake City, Utah 84112

S. H. ZERONIAN, Division of Textiles and Clothing, University of California, Davis, California 95616

Preface

Materials and energy from renewable resources: that concept has recaptured the American imagination. This resurgence of interest in biomass for a variety of uses can be related to a growing awareness of limitations associated with more finite mineral resources, particularly oil and natural gas. Increasing attention is being devoted now to the application of these renewable resources. Supply problems in areas such as energy, paper, chemicals, textiles, and structural materials can be addressed by renewable resources. Crops of cotton and conifer, for example, traditionally performed roles in supplying such materials. That they will be called upon for vastly expanded contributions is now being recognized.

The problems of converting this recognition—of a needed new role for carbohydrates and lignins—into action are many and diverse. They include increasing our knowledge concerning the specific characteristics of such materials. They also include the identification of specific means by which these materials might make contributions. Thus, the Cellulose and Paper and Textile Divisions of the American Chemical Society organized a joint symposium, held September 1–2, 1976, at San Francisco. The 16 papers of this volume were presented at that time. They represent the diversity and complexity of uses for renewable resources. They also demonstrate the variety of problems associated with increasing the contributions to society of these agricultural and silvicultural resources. On one end of the spectrum, controling forest fires—conflagrations that reduce the availability of such resources—gains treatment. On the other end, utilization of residues as a feedstock for the production of chemicals and fuels gains consideration. At one level, the papers present fundamental research; at another they consider specific processes for resource utilization. All, however, relate to the problem of turning cellulose and lignin into useful products for society.

The editors wish to express their appreciation to the contributors who made their research findings and papers available for this volume.

COMBUSTION CHARACTERISTICS OF CELLULOSIC FUELS

Fred Shafizadeh and William F. DeGroot

Wood Chemistry Laboratory
Department of Chemistry
University of Montana
Missoula, Montana

The readily available cellulosic materials, historic-
ally, have provided an abundance of natural fuel for mankind.
As ancient man spread out from his original habitat in tropi-
cal forests to occupy virtually all the earth, he seems to
have made campfires for protection and supplementary heat.
Use of fire as well as stone tools and language probably dis-
tinguished mankind from other primates by the beginning of
the Pleistocene Era. One of the most momentous events in the
development of mankind must have been the discovery of a
method for generating fire through the ignition of vegetation
or cellulosic materials.

Since the dawn of civilization, various types of wood
and cellulosic natural fuels have been consistently used for
heating, baking, making pottery and for refining of gold,
silver, and other metals. Also, partial combustion and

1

destructive distillation of wood have been used for making
charcoal, wood alcohol, tar, and other chemicals.

These applications, however, were recently phased out
when our rapidly developing industrial economy was geared to
the production of massive amounts of fossil fuel. Now that
we are facing severe limitations, both in the cost and availability of fossil fuels, it is only natural to reevaluate
wood and other cellulosic materials as an alternative fuel.

Utilization of the cellulosic fuels as a renewable
source of energy and their conservation by prevention of
hazardous fires require a knowledge of combustion characteristics of these materials that are closely related to their
chemical composition. This article provides a discussion of
the relationship between the chemical composition, combustion, heat release, and pyrolysis of cellulosic materials.

I. COMPOSITION

The wood residues contain lignocellulosic materials,
which form the plant cell walls, absorbed and condensed
moisture, various extractives and some mineral compounds.
These components could vary for different parts and species
of the plant. However, on a dry basis, the wood substance
contains about 7% of extractives and minor amounts of minerals or ash. The extracted hardwoods contain about 43% cellulose, 35% hemicelluloses, and 22% lignin, whereas the softwood contains about 43% cellulose, 28% hemicelluloses, and
29% lignin.

Acetyl-4-O-methylglucuronoxylan (xylan) forms the main
hemicellulose of the hardwoods, and glucomannans (mannan)
form the principal hemicelluloses of the softwoods. Furthermore, the softwood lignin contains guaiacyl propane units
(having one methoxyl group), and hardwood lignin, in addition
to this, contains syringyl propane units (with two methoxyl
groups). The higher content of acetyl and methoxyl groups in
hardwoods explains why this material has been used in
destructive distillation processes to obtain acetic acid and
methanol.

In comparison to the wood substance, bark contains
much more extractives and lignin or phenolic compounds, which

have a higher heat of combustion, and a lower amount of holocellulose (combination of cellulose and hemicellulose), which
has a lower heat of combustion. The needles and leaves also
have a higher percentage of extractives, as shown in Table I.

TABLE I

Analysis of Douglas Fir Wood, Bark, and Needles

Fraction	Content (%)		
	Wood	Bark	Needles
Ether-soluble	1.2	34.4	11.2
Benzene-alcohol-soluble	4.4	--	19.1
Hot water-soluble	5.6	29.2	7.2^a
Ash	0.2	0.9	5.6
Lignin and phenolics	27.2	26.1	36.4
Holocellulose	61.2	9.4^a	20.5

a. By difference.

II. COMBUSTION

The lower molecular weight ether extractives, particularly the terpenoid hydrocarbons (turpentine components),
have a high calorific value and readily evaporate and burn
in the gas phase with flaming combustion. The lignocellulosic materials, however, are not directly combustible, but
under the influence of a sufficiently strong source of energy
or pilot ignition, decompose to form volatile pyrolysis
products, which burn in the gas phase with flaming combustion.
The charred residue burns at a relatively slower rate by
surface oxidation or glowing combustion.

In these transformations, shown in Fig. 1, the cellulosic component is mainly converted to the combustible and
noncombustible volatiles, including water and carbon dioxide,
whereas the lignin component contributes mainly to the char
fraction. The initial depolymerization of cellulose produces an intermediate tar fraction containing levoglucosan.

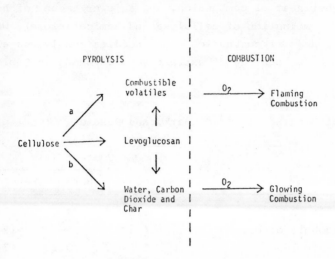

Fig. 1. Competing reactions in the pyrolysis and combustion of cellulose.

The rate of gasification and the amount of residue left at different temperatures could be determined by thermal analysis, as shown for cellulose (Fig. 2). Thermal analysis of cottonwood and its components, shown in Figs. 3 and 4, indicate that the pyrolysis of wood reflects the thermal properties of its components.

At this stage, it should be pointed out that char formation is also a function of the rate of heating, the particle size, and the presence of moisture and inorganics. Higher temperature, faster heating rate, and smaller particle size promote the gasification process and the rapid flaming combustion, whereas lower temperatures, larger particle size, and the presence of moisture and inorganics increase the amount of char and favor the smoldering combustion or no combustion at all.

In this connection, it should also be noted that the hydrocarbons and lipids classified as the ether extractives not only burn more rapidly and generate more heat than the wood substance, but they also promote the rapid and more

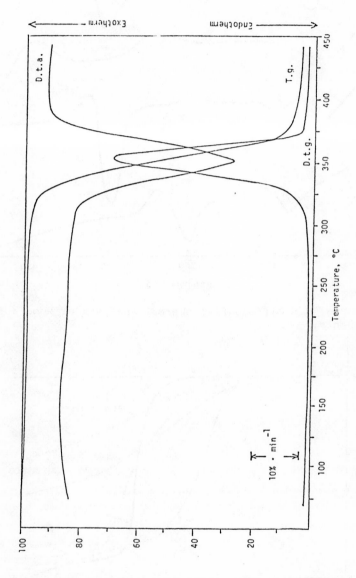

Fig. 2. Thermal analysis curves of microcrystalline cellulose.

Fred Shafizadeh and William F. DeGroot

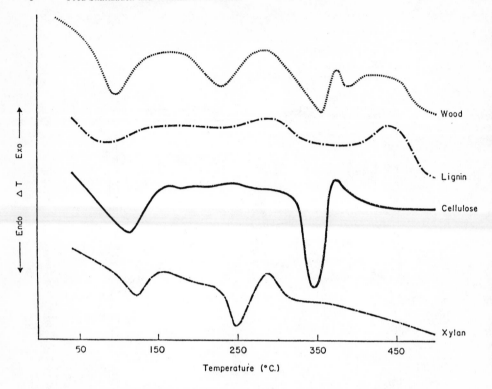

Fig. 3.   Differential thermal analysis of wood and its components.

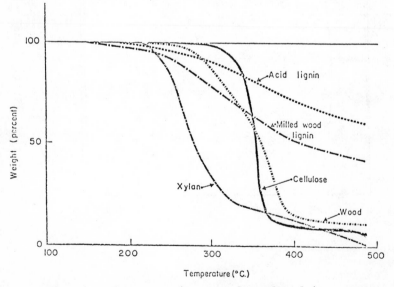

Fig. 4.   Thermogravimetry of wood and its components.

complete combustion of the lignocellulosic material by
increasing the temperature and the rate of gasification.

III. HEAT RELEASE

The heat contents of different types of forest fuels
and their components are given in Table II. This table also
shows the amount of char left at 400°C on heating at the rate
of 200°C/min and the heat of combustion of the gaseous and
carbonaceous products. Table III shows the distribution of
the heat of combustion between the char and the volatile
products. These data show the relationship between the heat
of combustion and the chemical composition of the fuel.

Pure cellulose, which is composed entirely of sugar
units, with the elemental analysis of $C_6(H_2O)_5$, has a relatively low heat content because of the high level of oxidation or water content. However, 75% of its heat content is
released to the volatiles because of the transglycosylation
and thermal cleavage of the polysaccharide, discussed later.

TABLE II

Heat of Combustion of Natural Fuels and Their Pyrolysis Products as Char and
Combustible Volatiles

Source	Type	Fuel $\Delta H^{25°}$ comb (cal/gm)	Char Yield (%)	Char $\Delta H^{25°}$ comb (cal/gm)	Combustible Volatiles Yield[a] (%)	Combustible Volatiles $\Delta H^{25°}$ comb (cal/gm)
Cellulose	Filter paper	-4143	14.9	-7052	85.1	-3634
Poplar wood (Populus spp.)	Excelsior	-4618	21.7	-7124	78.3	-3923
Larch wood (Larix occidentalis)	Heart wood	-4650	26.7	-7169	73.3	-3732
Decomposed Douglas fir (Pseudotsuga menzeisii)	Punky wood	-5120	41.8	-7044	58.2	-3738
Ponderosa pine (Pinus ponderosa)	Needles	-5145	37.0	-6588	63.0	-4298
Aspen (Populus tremuloides)	Foliage	-5034	37.8	-6344	62.2	-4238
Douglas fir lignin	Klason	-6371	59.0	-7416	41.0	-4867
Douglas fir bark (Pseudotsuga menzeisii)	Outer (dead)	-5122	52.8	-5798	47.2	-4366
	Whole	-5708	47.1	-6406	52.9	-5087

a. Heating rate 200°C/min.

TABLE III.

Distribution of the Heat of Combustion of Forest Fuels

Fuel		Char (cal/gm fuel)	Gas (cal/gm fuel)	Total (cal/gm)
Source	Type			
Cellulose	Filter paper	-1050	-3093	-4143
Poplar wood (Populus spp.)	Excelsior	-1546	-3072	-4618
Larch wood (Larix occidentalis)	Heart wood	-1914	-2736	-4650
Decomposed Douglas fir (Pseudotsuga menzeisii)	Punky wood	-2944	-2176	-5120
Ponderosa pine (Pinus ponderosa)	Needles	-2438	-2708	-5146
Aspen (Populus tremuloides)	Foliage	-2398	-2636	-5034
Douglas fir lignin	Klason	-4375	-1995	-6370
Douglas fir bark (Pseudotsuga menzeisii)	Outer (dead)	-3061	-2061	-5122
	Whole	-3017	-2691	-5708

Excelsior (poplar) and larch woods, in addition to the cell wall polysaccharides, contain about 25% of lignin. This macromolecule is composed of coniferyl units (and some related syringyl units in poplar), with the elemental analysis of $C_{10}H_{11}O_2$, representing a lower degree of oxidation and a higher heat of combustion than that obtained for cellulose (6371 versus 4143 cal/gm). Also, it is not readily cleaved by pyrolysis and contributes mainly to the char. Consequently, the excelsior and larch samples have higher heats of combustion, but lower fractions of their total heat content (65% and 58%, respectively) are found in the volatiles.

The effect of the lignin content is more pronounced with the punky wood samples in which the cell wall polysaccharides are partially removed by biological degradation resulting in 52% lignin content. This sample has a still higher heat of combustion but lower heat release ratio of 41%. This is why the punky wood supports smoldering or latent combustion, which holds the fire over long periods of time, rather than the rapidly developing and consuming flaming combustion.

The effect of ether extractives (terpenoid hydrocarbons and lipids), which have a still lower oxygen and higher heat content ($\sim$7700-8500 cal/gm) on samples of ponderosa pine and aspen foliage, is also interesting, but complicated by the presence of some ash and various amounts of lignin in addition to the ether extractives. However, comparison with the corresponding extracted samples showed that the absence of ether extractives, not only lowers the total heat of combustion of the fuel, but also lowers the ratio of heat release from 50% to 42% for aspen foliage. In other words, the extractives provide a source of readily vaporized and high-energy combustible materials. These considerations indicate a definite correlation between the thermal properties of natural fuels and their chemical composition both at the molecular and atomic levels.

At the latter level, the heat content is clearly related to the oxidation state of the natural fuels in which carbon atoms generally dominate and overshadow small variations of hydrogen content. This situation leads to a rather unexpected, but highly interesting and useful correlation. When the heats of combustion of various substrates, including the fuels, chars, and volatiles, are plotted as a function of their respective carbon content (Fig. 5), the least squares line through the individual points fits the equation

$$\triangle H^{25°} \text{ combustion (cal/gm)} = 94.19(\%C) + 55.01.$$

This correlation is due to an averaging and mutual cancellation of different effects, particularly the effect of hydrogen and oxygen contents. Hydrogen atoms should increase the heat content above the average value and oxygen atoms should reduce it below the average value. For instance, in the extreme case of carbon dioxide, the heat of combustion is nil and the carbon content is 27.3%. However, within the range of chars, fuels, and volatiles that have been considered, the chars have lower hydrogen content, but also lower oxygen content; the reverse is true for the volatiles and the original fuels that are in between. In all of these substrates, the two opposing effects cancel each other.

For a realistic evaluation of forest fuel, the heats of combustion obtained from calorimetric determination at 25°C should be corrected to the ignition temperature by

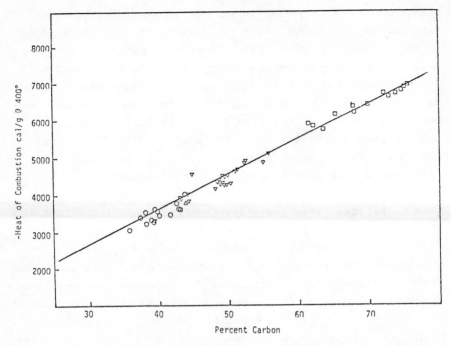

Fig. 5.   Heat of combustion at 400°C versus percent carbon.   (▽) fuels, (□) char, (o) volatiles.

considering the heat of preignition required for heating the forest fuels to the ignition temperature. The heat of pre-ignition includes the energy required for vaporization of the moisture content (about 582 cal/gm), the heat of pyroly-sis, and the heat capacities required for raising the temp-erature of the various products (200-250 cal/gm).

The balance between the heat of combustion of the volatile products and the heat of preignition provides a good indication of the flammability. These values could be obtained as a function of temperature by the thermal evolu-tion analysis (TEA), which gives the heat of combustion of the volatiles, and differential scanning calorimeter (DSC), which gives the heat of preignition.

Figure 6 shows the heat of combustion of the volatiles produced by gasification of Douglas fir needles at different temperatures before and after removal of the extractives. This figure and similar data obtained for various forest

fuels dramatically demonstrate the contribution of the
extractives to the combustibility of forest fuels.

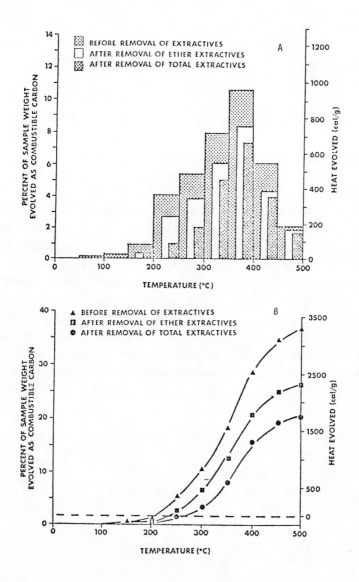

Fig. 6. Evolution of carbon and heat from Douglas
fir foliage, (A) in temperature intervals and (B) cumulative,
based on dry weight of the unextracted sample.

IV.   PYROLYSIS OF CELLULOSIC MATERIALS

Thermal degradation of cellulose and hemicelluloses to flammable volatile products and chars involves a series of highly complex reactions and a variety of products, which have been extensively investigated in the author's laboratory. These reactions, which take place both concurrently and consecutively, may be classified in the following categories:

a.   Depolymerization of the polysaccharides by trans-glycosylation at about 300°C to provide a mixture of levoglucosan, other monosaccharide derivatives, and a variety of randomly linked oligosaccharides, as shown in Fig. 7 and Table IV for cellulose. This mixture is generally referred to as the tar fraction.

b.   The above reactions are accompanied by dehydration of sugar units in cellulose. These give unsaturated compounds, including 3-deoxyglucosenone, levoglucosenone, furfural, and a variety of furan derivatives which are found partly in the tar fraction and partly among the volatiles.

c.   At somewhat higher temperatures, fission of sugar units provides a variety of carbonyl compounds, such as actaldehyde, glyoxal, and acrolein, which readily evaporate.

d.   Condensation of the unsaturated products and cleavage of the side chains through the free radical mechanisms leaves a highly reactive carbonaceous residue containing trapped free radicals.

Heating of the cellulosic materials at or above 500°C provides a mixture of all of these products as shown in Tables V and VI for cellulose and xylan.

Addition of an acidic catalyst or slow heating promotes the dehydration and charring reactions, charcoal formation, and glowing ignition, whereas rapid heating to high temperatures promotes the fission reactions, volatilization, and flaming combustion.

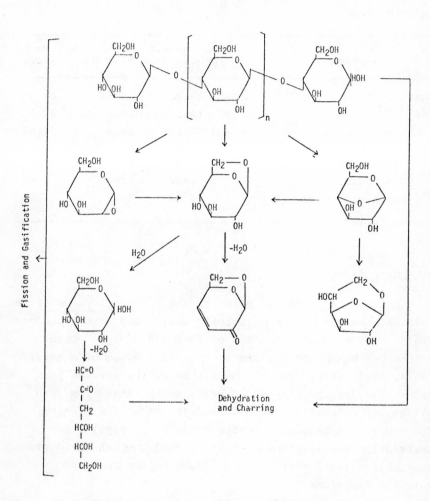

Fig. 7. Thermal degradation of cellulose.

TABLE IV

Analysis of the Pyrolysis Products of Cellulose at 300°C under Nitrogen

Condition	Atmospheric pressure	1.5 Mm Hg	1.5 Mm Hg, 5% SbCl$_3$
Char	34.2%	17.8%	25.8%
Tar	19.1	55.8	32.5
levoglucosan	3.57	28.1	6.68
1,6-anhydro-β-D-glucofuranose	0.38	5.6	0.91
D-glucose	trace	trace	2.68
hydrolyzable materials	6.08	20.9	11.8

V. CONCLUSIONS

The data presented point to the potential value of the colossal amounts of the cellulosic wastes and forest product residues as renewable sources of fuel and energy. On a weight per weight basis, the average heat value of natural fuels, 4500 cal/gm, is not too far below the average heat values for different types of coal, which range from 7915 cal/gm for bituminous grade A to 3640 cal/gm for lignite. The comparison becomes more favorable by pyrolysis or destructive distillation of natural fuels, which could provide charcoal and various chemicals from the char and volatile fractions.

The new scientific information and technology could be applied to the gasification of wood to provide a number of products and by-products, ranging from solid, liquid, and gaseous fuel to activated charcoal and polyfunctional compounds as a substitute for the petrochemicals.

TABLE V

Pyrolysis Products of Cellulose and Treated Cellulose at 550°C

	Yield (%)[a]			
Product	Neat	+5% H_3PO_4	+5% $(NH_4)_2HPO_4$	+5% $ZnCl_2$
Acetaldehyde	1.5	0.9	0.4	1.0
Furan	0.7	0.7	0.5	3.2
Propenal	0.8	0.4	0.2	Trace
Methanol	1.1	0.7	0.9	0.5
2-Methylfuran	Trace	0.5	0.5	2.1
2,3-Butanedione	2.0	2.0	1.6	1.2
1-Hydroxy-2-propanone; Glyoxal	2.8	0.2	Trace	0.4
Acetic acid	1.0	1.0	0.9	0.8
2-Furaldehyde	1.3	1.3	1.3	2.1
5-Methyl-2-furaldehyde	0.5	1.1	1.0	0.3
Carbon dioxide	6	5	6	3
Water	11	21	26	23
Char	5	24	35	31
Tar	66	16	7	31

a. Yield based on the weight of the sample.

TABLE VI

Pyrolysis Products of Xylan and Treated Xylan at 500°C

Product	Yield (%)[a]			
	Xylan		O-Acetylxylan	
	Neat	+10% $ZnCl_2$	Neat	+10% $ZnCl_2$
Acetaldehyde	2.4	0.1	1.0	1.9
Furan	Trace	2.0	2.2	3.5
Acetone; propionaldehyde	0.3	Trace	1.4	Trace
Methanol	1.3	1.0	1.0	1.0
2,3-Butanedione	Trace	Trace	Trace	Trace
1-Hydroxy-2-propanone	0.4	Trace	0.5	Trace
3-Hydroxy-2-butanone	0.6	Trace	0.6	Trace
Acetic acid	1.5	Trace	10.3	9.3
2-Furaldehyde	4.5	10.4	2.2	5.0
Carbon dioxide	8	7	8	6
Water	7	21	14	15
Char	10	26	10	23
Balance (tar)	64	32	49	35

a. Yield based on the weight of the sample.

REFERENCES

1. F. Shafizadeh, "Pyrolysis and Combustion of Cellulosic Materials," Advan. Carbohyd. Chem. 23, 419 (1968).

2. F. Shafizadeh, "Cellulose Chemistry: Perspective and Retrospect," Pure Appl. Chem. 35, 195 (1973).

3. F. Shafizadeh, "Industrial Pyrolysis of Cellulosic Materials," Appl. Polym. Symp. 28, 153 (1975).

4. F. Shafizadeh, "Thermal Behavior of Carbohydrates," J. Polym. Sci., Part C, 36, 21 (1971).

5. R. A. Susott, W. F. DeGroot, and F. Shafizadeh, "Heat Content of Natural Fuels," J. Fire Flammability 6, 311 (1975).

KINETICS OF SOLID-PHASE CELLULOSE PYROLYSIS

A. Broido

Pacific Southwest Forest and Range Experiment Station
Forest Service, U.S. Department of Agriculture
Berkeley, California

When the course of a pyrolysis is determined by measuring weight change with time, it is possible to derive kinetic data for prior reaction steps in which no weight change occurs. Recognition of this fact has led to determination of rate constants and empirical formulas for each step in a previously-proposed pyrolysis scheme in which an initial "incubation period" with no weight loss is followed by branching into (a) a depolymerization leading to completely volatile products and (b) a sequence of steps leading to char formation. The rate constant (hr^{-1}) for depolymerization may be expressed as

$$K_B = 6.11 \times 10^{18} \exp(-52{,}940/RT).$$

The first step in the char-forming sequence involves no weight loss, and its rate constant (hr^{-1}) is given, somewhat less reliably, by

$$K_2 = 5.75 \times 10^{13} \exp(-40,500/RT).$$

For reasonable times at low temperatures, the last step in the char-forming sequence involves a weight loss corresponding to formation of one $C_8H_6O_2$ per repeating cellobiose unit in the original cellulose. For longer times at higher temperatures or for cellulose contaminated with inorganic catalysts, an additional step leads to $(C_7H_4O)_n$. It appears that the inorganic impurities have little direct effect on K_B but markedly increase K_2.

I. INTRODUCTION

The ability to detect, or even define, individual steps in a sequence of chemical reactions such as occur in a pyrolysis process depends, inter alia, upon the measurement techniques. Thus, in following the weight change in the pyrolysis process, a relatively slow rate-determining reaction followed by a sequence of much faster reactions are all lumped into a single "step." The reaction rate for such a step is the rate of the slow reaction, but the weight change is the overall weight loss of the entire sequence of reactions in the step. If the course of the pyrolysis is followed by measuring weight change as a function of time, i.e., by thermogravimetry, a fallacy generally accepted as a truism holds that no kinetic data can be obtained for reaction steps in which no weight change occurs.

In 1971, an attempt was made to resolve into the minimum number of consecutive and competing reactions a continuously recorded weight-loss curve obtained in a 1000 hour (6 week) vacuum pyrolysis of high-purity cellulose paper at 226°C [1]. To account for an initial "incubation" period with no weight loss, as well as the previously established branching into (a) a depolymerization leading to completely volatile products and (b) a sequence of weight-loss steps leading to char formation, the model adopted took the form shown in the following diagram:

$$B' \quad \text{(volatile tars)}$$

$$
\begin{array}{c}
B' \quad \text{(volatile tars)} \\
\uparrow K_B \\
A \xrightarrow{K_1} B \xrightarrow{K_2} C \xrightarrow{K_3} D \xrightarrow{K_4} E \\
\text{wt} = 1 \quad \text{wt} = 1 \quad \text{wt} = w_c \quad \text{wt} = w_d \quad \text{wt} = w_e
\end{array}
$$

(char formation)

where C. D. and E represent only the solid products of steps 2, 3, and 4, and any volatile fraction evolved in one of these steps would be manifested as a reduction in the relative weight of the corresponding solid product.

Two solutions were found that fit the entire curve within experimental error (the number of solutions goes up rapidly as the allowable error is increased). Both solutions, however, led to conclusions in conflict with other observations. Further, attempts to use logically consistent extensions of either of these solutions to resolve subsequent isothermal experiments at higher temperatures were unsuccessful.

The recent phasing out of this research program provided an opportunity for a final attempt to obtain a model consistent with all the available data. This attempt led to the recognition that cross-linking and rearrangement reactions involving no weight loss [2], previously assumed to be rapid sequences included within a weight-loss step, could, in fact, be represented as complete steps in the above scheme, i.e., that weight measurements could be used to provide kinetic data for prior reaction steps with no weight loss. This paper describes such a model, illustrates its consistency with the data at 226°C and uses the observations at higher temperature in the computation of pertinent Arrhenius equations.

II.   EXPERIMENTAL WEIGHT LOSS DATA

The cellulose samples used in these experiments, about 3 x 5 cm in cross section and weighing about 100 mg, were cut from Whatman No. 541 "ashless" filter paper (ash content 0.008%). The thermogravimetry apparatus for the experiment

at 226°C has been described previously [1]. The slow reaction rate at temperatures below 226°C effectively precluded
additional experiments at such low temperatures. With the
higher reaction rates at temperatures 20-30°C higher (Fig. 1),
not only did replication become feasible, but each experiment could be extended to more nearly approach "infinite"
time.

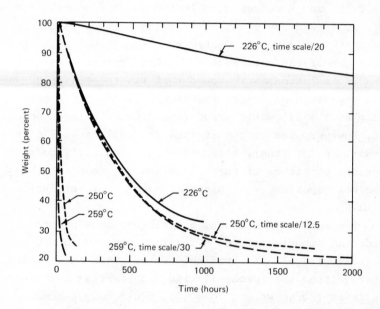

Fig. 1. Residual weight of cellulose sample versus time
at pyrolysis temperature.

For the experiments at temperatures in the 245-260°C
range, both weight and temperature were logged continuously
on a two-pen strip chart recorder. These experiments were
plagued with the usual experimental complications--leaks in
the vacuum system, electrostatic charges (especially when the
operator was wearing synthetic laboratory coats) causing the
sample to stick to the sides of the reaction tube or to brush
against the thermocouple, temperature excursions or recorder
noise well beyond the usual range. The dozen experiments
that were chosen for analysis experienced a minimum of such
complications. Nevertheless, while the relative error in the

weight measurement at 226°C was previously estimated to be
less than 0.3%, for the later experiments, recorder noise and
occasional brief "sticking" sometimes doubled this error. A
more serious problem, however, clearly limiting the upper
temperature at which accurate data could be obtained, was
revealed by the continuous temperature recording.

Even with manual manipulation of the furnace controls,
it took some 15-20 min to reach (i.e., $\pm 1°C$) isothermal
conditions. For the experiment at 226°C, weight loss during
this heat-up time was negligible, but as temperatures
approached 260°C, measurable reaction occurred during heat-
up. After the system became isothermal, temperature excursions as large as $\pm 1°C$ and lasting several hours were
recorded. For activation energies near 50 kcal/mole, a 1°C
temperature change causes a 10% change in reaction rate.
Again, an effect of this magnitude for a few hours is trivial
for reactions progressing as slowly as indicated by the 226°C
curve, but at 260°C, the effect would be expected to produce
a noticeable change in the curve.

With the apparatus available, then, 260°C served as an
upper limit to the temperature range of these experiments.
The reductions in time scale at the higher temperatures
meant that an experiment equivalent to 1000 hr at 226°C could
be completed in about 80 hr at 250°C and in less than 36 hr
at 260°C.

III. FITTING OF 226°C WEIGHT-LOSS CURVE

As before [1], the computerized curve-fitting procedure was used to obtain empirically the minimum number of
reaction steps required for any predetermined degree of fit,
together with the values of rate constants and normalized
molecular weights required for the particular scheme. However, the present search included steps with no weight loss.
Surprisingly, a solution to the previous scheme was found in
which $w_c = w_d = 1$, and, over the duration of the curve, all
volatilization in the char-forming sequence occurs in a
single final step, with $w_e = 0.4136$ (i.e., the steps
$A \longrightarrow B \longrightarrow C \longrightarrow D$ each occur without weight loss, and all loss

prior to steps D ⟶ E results from the tar-forming reaction,
B ⟶ B'). For values of the rate constants as listed in the
heading to Table I, the maximum deviation between theoretical
and experimental weights was < 0.25%. The representative
times given in Table I indicate the peak weights for the
various components, as well as the maximum deviations between
the computed and experimental results.

TABLE I

Residual Weight of Cellulose Sample after Various Times at 226°C[a]

Time (hr)	Computed weight of							Expt'l weight
	A	B	B'[b]	C	D	E	Sum[c]	
0	100.00	0.00	(0.00)	0.00	0.00	0.00	100.00	100.00
20	2.73	84.83	(3.46)	8.68	0.29	0.00	96.53	96.78
50	0.01	66.60	(9.28)	21.86	2.03	0.09	90.59	90.35
150	0.00	26.63	(20.40)	37.44	11.17	1.80	77.04	77.29
350	0.00	4.25	(26.63)	26.20	18.14	10.25	58.84	58.81
500	0.00	1.08	(27.51)	15.74	14.98	16.83	48.63	48.47
750	0.00	0.11	(27.78)	6.00	7.74	24.14	37.99	37.86
1000	0.00	0.01	(27.81)	2.18	3.32	27.58	33.09	33.33

    a. Comparison of experimental curve with values computed for $w_a = w_b = w_c = w_d = 1$, $w_e = 0.4136$, $w_f = 0.3210$, and the following values of $K \times 10^3$/hr: $K_B = 2.55$, $K_1 = 180$, $K_2 = 6.62$, $K_3 = 4.11$, $K_4 = 6.30$, and $K_5 = 0$. Representative times indicate peak values for the various components and maximum deviations between computed and experimental results.

    b. Completely volatile; does not contribute to residue sum.

    c. Maximum deviation from experimental weight = 0.247.

The value for w_e is appreciably higher than those
found in the earlier solutions and corresponds to a residue
with the empirical formula C_4H_3O per anhydroglucose monomer
unit of the original cellulose. Since no such compound with
an odd number of hydrogens is possible, the value implies a
cellobiose dimer as the repeating unit in the pyrolysis
process. The weight loss for each celloboise unit corresponds to the loss of CO, CO_2, $3H_2O$, and $2CH_3OH$ as such or in
equivalent combinations (e.g., CH_3COOH instead of
$CH_3OH + CO$).

The elemental composition of the total theoretical residue at 1000 hr may be computed as C = 67.00%, H = 4.78%, and O = 28.20%. The accuracy attainable in elemental analysis of the experimental residue is limited by two major complications: (a) combustion analysis determines carbon and hydrogen, with oxygen obtained only by difference, and (2) correction is required for absorption of atmospheric gases, primarily water. Consequently, the remarkable agreement of the above calculated values with the experimental values C = 67.25%, H = 4.73%, and O = 28.02% should be considered somewhat fortuitous.

One major drawback to the two earlier solutions had been that the values for w_e (either at or close to a pure carbon char) seemed to preclude the roughly 35% additional weight loss observed when the residue after 1000 hr at 226°C was heated to 370°C for an additional 2 hr. If w_e = 0.4136, the weight loss at 370°C corresponds well with the reaction of all the components of the 1000 hour residue (Table I) through the following additional step in the pyrolysis sequence:

$$(C_8H_6O_2)_n \xrightarrow{\quad K_5 \quad} (C_7H_4O)_n + nCH_2O\uparrow$$

Chemical analysis of the char remaining after this supplemental heating gave C = 80.87%, H = 3.92%, and O = 15.21%, again in remarkable agreement with the values C = 80.77%, H = 3.85%, and O = 15.38% for the empirical formula C_7H_4O. We have no other evidence for the formula C_7H_4O--such close agreement with any simple empirical formula would seem to be remarkable. For example, the earlier solutions both required minor deviations from exact empirical formulas for the various weight-loss steps (e.g., w_c = 0.893, where the loss of one H_2O per anhydroglucose unit would require 0.889, or w_e = 0.355, where an empirical formula of $C_4H_2O._5$ corresponds to 0.358).

With the weight changes in each reaction step fixed, each experimental curve was fitted by selecting appropriate values for the five rate constants (six, if K_5 is also included). It might appear that these rate constants provide sufficient degrees of freedom to permit almost any curve to

be fitted. It should be noted, however, that not all constants have a significant effect on all parts of the curve.
Thus, (Table II) values of each rate constant in Table I may
be changed in turn to produce a 1% weight change at 500 hr--
the time midpoint of the experiment. Note that only changes
in K_1 and K_B influence the beginning portion. Again, with
this scheme only K_B and K_2 can affect the residue at infinite
time. In fact, K_B is the only constant affecting all portions of the curve, and thus is the constant that can be
determined with greatest accuracy. Another noteworthy point
is that, in contrast with changes in any of the other constants, increasing K_2 decreases the overall weight loss rate,
since it decreases the fraction of the sample losing weight
by depolymerization.

IV. ARRHENIUS EQUATION FOR DEPOLYMERIZATION

An earlier dynamic thermogravimetry experiment [3]
provided an estimate of E_B, the activation energy for the
depolymerization process, as 54.9 $\pm$ 2 kcal/mole, with pre-
expenential factor a function of the exact value of E_B
chosen. That experiment provides its most accurate estimate
of K_B at the temperature of maximum weight change, 328°C.
That value can be combined with the present value of K_B at
226°C to yield the following Arrhenius expression:

$$K_B = 6.11 \times 10^{18} \exp(-52,940/RT) \qquad (1)$$

Although the introduction of K_5 is not needed to fit
the 226°C curve within experimental error, an even closer fit
would be expected if K_5 is permitted to have a non-zero value.
For such a solution (Table III), the value of K_B is some 15%
higher than when $K_5 = 0$. Of course, agreement with the
elemental analysis and ultimate weight loss has fallen drastically (the theoretical composition at 1000 hr is C = 69.27%,
H = 4.62% and O - 26.11%). Even so, such a difference would
only lower the activation energy for depolymerization from
52.94 to 52.10 kcal/mole.

With the sequence of steps and the weight change in
each now fixed, it was found possible to fit the other

TABLE II

Influence of Changes in Rate Constants on Pyrolysis Curve[a]

Time (hr)	Expt'l weight	Theoretical weight, K values as for Table I, except:						
		None	K_B=0.00239	K_1=0.0455	K_2=0.00759	K_3=0.00378	K_4=0.00565	K_5=0.00168
20	96.78	96.54	96.75	98.35	96.56	96.54	96.54	96.54
50	90.35	90.59	91.16	93.42	90.76	90.60	90.61	90.59
150	77.29	77.04	78.15	78.99	77.83	77.22	77.25	77.02
350	58.81	58.84	60.00	60.16	59.93	59.61	59.68	58.44
500	48.47	48.63	49.63	49.63	49.63	49.63	49.63	47.63
750	37.86	37.99	38.76	38.50	38.96	38.89	38.77	35.71
1000	33.33	33.09	33.73	33.31	34.11	33.68	33.53	29.58
Infinity[b]	23.29	23.17	23.59	23.17	24.03	23.17	23.17	23.17
Maximum deviation[c]		0.247	1.27	3.07	1.24	1.21	1.19	3.75
Time of maximum deviation (hr):		150	400	50	450	600	450	1000

a. Chosen values of K produce a change of 1.00 mg at 500 hr.

b. Experimental weight as T→infinity is weight of residue after 2 hr at 370°C; theoretical weight is computed as 32.1 $K_2/(K_B + K_2)$.

c. Maximum deviation from the experimental weight.

TABLE III

Representative Rate Constants from Curve Fitting with and without Nonzero Values for K_5

Temp. (°C)	$10^2 K_B$ (hr^{-1})	K_1 (hr^{-1})	$10^2 K_2$ (hr^{-1})	$10^2 K_3$ (hr^{-1})	$10^2 K_4$ (hr^{-1})	$10^3 K_5$ (hr^{-1})
226	0.255	0.180	0.662	0.441	0.630	0
	0.259	0.171	0.699	0.507	0.507	0
	0.293	0.120	1.00	0.390	0.774	0.042
248	(2.19)[a]		(3.44)[b]			0
	2.46	1.17	6.06	8.28	3.51	infinity
250	2.70	1.64	4.17	6.27	6.66	0
	3.05	1.17	7.26	5.82	8.40	1.08
259	6.83	3.38	7.94	1.15	1.57	0
	7.27	2.79	10.82	1.60	1.31	1.30

a. Computed from $K_B = 6.11 \times 10^{18} \exp(-52{,}940/RT)$.
b. Computed from $K_2 = 5.75 \times 10^{13} \exp(-40{,}500/RT)$.

isothermal experiments within experimental error by appropriate changes in the various rate constants. To see how Eq. (1) agrees with the data at higher temperatures, a conventional procedure would be to estimate, for each experiment, the range of permissible values of K_B at "the" temperature of the experiment. While this range differed among the experiments (depending on the quality of the data), temperature fluctuations within each experiment were within ± 1°C. Consequently, the somewhat unconventional plotting procedure chosen (Fig. 2) was to accept the determined value of K_B as applying somewhere within the ± 1°C temperature interval. For the best (in the sense of appearance of the original recorder tracing) experiment near 250°C and 260°C, two fits were made (Table III and Fig. 2)--with and without the inclusion of K_5. Despite differing degrees of uncertainty in the estimated values of K_B for the experiments, none of the points deviates from the curve by more than 1°C.

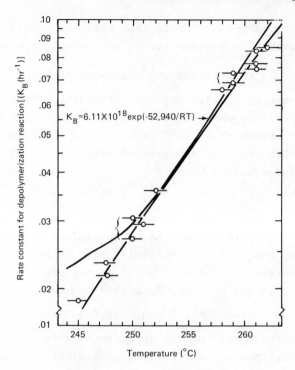

Fig. 2.   Reaction rate for depolymerization versus pyrolysis temperature. Bracketed pairs of points each represent two fits, with and without inclusion of K_5, to the same experimental curve.

V.   RATIO OF DEPOLYMERIZATION TO CHAR-FORMING PROCESSES

Since K_B and K_2 alone determine the fraction of the cellulose sample pyrolyzing by either the depolymerization or the char-forming sequences, these two constants are of overriding importance in determining the combustion behavior of a cellulose sample. Unfortunately, neither the accuracy in determining K_2 for a given experiment nor its reproducibility among experiments can be as good as for K_B.

The reduced accuracy in curve fitting for a single experiment is evident from the fits with and without K_5 (Table III). For the experiment at 226°C, for example, the nonzero value for K_5 (corresponding to 15% of the sample having reacted to a residue of 0.3210 by 1000 hr) changes K_B by

only 15% but changes K_2 by more than 50%. Nevertheless, the
solutions with $K_5 = 0$ given for K_2 the Arrhenius expression

$$K_2 = 5.75 \times 10^{13} \exp(-40,500/RT) \qquad (2)$$

and this equation yields, for the previously defined "best"
higher-temperature experiments, computed temperatures of
250.1°C (experimental = 250°C) and 258.9°C (experimental
= 259°C). For the same experiments with nonzero values of K_5,
the Arrhenius equation becomes

$$K_2 = 1.94 \times 10^{14} \exp(41,300/RT) \qquad (3)$$

and the two computed temperatures are 251.5 and 257°C. As a
worst case, if K_5 is assumed to be negligible at 226°C (to
conform with the elemental analysis) and nonnegligible at the
higher temperatures, the activation energy more nearly
approaches that for K_B. The Arrhenius equation becomes

$$K_2 = 3 \times 10^{17} \exp(-49,000/RT) \qquad (4)$$

Only about half of the experiments were used in the
above computations for K_2; for the others, no solution within
experimental error could be found without invoking K_5. Even
before an attempt was made to fit the curves, simple comparison of their long-time tails indicated the existence of inter-
sample discrepancies. Thus, with $K_5 = 0$ and $w_e = 0.4136$,
Eq. (1) and (2) can be used to compute values in the range
22–25% as infinite-time residues for samples at temperatures
of 250–260°C, consistent with the experimental curves used in
the determination of Eq. (2). Those experiments for which no
solution could be found without use of K_5 had "asymptotic"
residues below the values so computed.

Contamination of pure cellulose with trace quantities
of inorganic impurities (e.g., ash constituents, fire retardants) is known to catalyze the char-forming processes. Consequently, high values of K_5 and the corresponding increase
in K_2 found in these solutions imply contamination of some of
the samples during handling or within the reaction chamber
(e.g., by inadvertent contact with the internal thermocouples). At first glance, finding lower residues for the
presumably contaminated samples seems to contradict the

expectation of increased char for such samples. Note, however, that while an increased value for K_5 only increases the
rate of weight loss in the char-forming sequence, an increased
value for K_2 also increases the fraction of the total sample
pyrolyzing by this sequence, thus inhibiting the greater
weight loss that would otherwise occur by depolymerization.
Therefore, for contaminated samples the rate of approach to
an infinite-time char weight would be faster, but the
infinite-time weight itself [which can be computed for this
model as $0.3210\ K_2/(K_B + K_2)$] would be higher. To whatever
extent the results do indicate a real catalytic effect, its
influence on K_5 appears to be much greater than on K_2. As an
extreme example (Table III), one solution was found with K_5
infinite ($\gg 10^{-2}$/hr). Even here, K_2 was only 1.75 times the
value computed from Eq. (2).

Further evidence that this extreme solution was not
just a computational aberration was obtained pragmatically.
If the residues of these low-temperature pyrolyses are heated
to 370°C for 2 hr and small samples ($<1\ cm^2$) of the chars so
formed are held vertically above a microtorch flame (1-2 mm
diameter), the leading edges of the char burn in a glowing
process that ends promptly as the sample is removed from the
flame zone. On the other hand, theoretical computations [4]
indicate that the gases evolved in steps 4 and 5 of the pyrolysis should, under appropriate circumstances, be combustible.
In general, when residue samples were held above the torch
flame, a feeble yellow flame formed and burned to the top of
the sample but with no lateral spread. However, no such
flame was observed with the residue for which K_5 was
infinite.

As was done for the experiment at 226°C, elemental
analysis of the residues on termination of the higher-
temperature experiments, accurate determination of the char
weights at infinite time (as obtained by heating the residues
to 370°C for 2 hr) and elemental analysis of these chars
could have resolved some of the remaining questions. Unfortunately, fiscal limitations precluded obtaining such data
when the experiments were first performed. The residues were
stored, and analysis of some of the freshly dried samples,

before and after heating to 370°C, was performed recently, as much as 4 years after the initial experiments. However, during the interim, the residues had gained weight (beyond simple water absorption). Both the elemental analyses and mass spectrometry on products evolved on further heating indicated that partial oxidation had occurred.

Mass spectrometry measurements were undertaken in an attempt to confirm H_2CO as the evolved product in the last pyrolysis step. A residue sample was mounted in a tube attached to the spectrometer vacuum system and the system was evacuated. The tube was then slowly heated (with pumping) to 210-220°C, the vacuum turned off, and an overnight "outgassing" sample collected at this temperature. Immediately after running this sample, we turned up the heater and again turned on the vacuum system. After a temperature increased another 100°C, a sample of the evolving pyrolysis gases was withdrawn for analysis. Uncorrected peak heights were measured relative to the mass 18 peak.

For all measurements, the major components were clearly H_2O, CO, and CO_2. For the outgassing sample, however, a smaller but significant N_2 peak was also found. A peak at mass 28, of course, includes both N_2 and CO, but a peak at 14 without correspondingly high peaks at 13 and 15 is strongly indicative of N_2. * No corresponding peak was detected for O_2, and this is taken to mean that, as air was absorbed by the sample, the O_2 reacted with the sample to yield a partially oxidized char. The height of the peak at mass 30, much smaller than those at 28 and 29 and thus taken as a (nonexclusive) indicator for H_2CO, was 0.4% that at mass 18. With pyrolysis at 310°C, the CO and CO_2 concentrations increased fivefold relative to the mass 18 H_2O peak. The peak at mass 30, while still small compared with the major peaks, increased by a factor of 20. While this result does not prove the existence of H_2CO as a pyrolysis product in these samples--much less in similarly prepared fresh samples--it at least does not exclude such a possibility.

---

* Ultramicro elemental analysis was carried out on a single residue sample and revealed 0.5% nitrogen.

Five of the stored residues were heated for 2 hr at 370°C to produce char residues for elemental analysis. In each case, the weight loss was appreciably greater than the corresponding loss from fresh residue after the experiment at 226°C. For three of the samples, elemental analysis showed an oxygen content higher than for C_7H_4O. Surprisingly, however, two of the chars gave results in excellent correspondence with such a formula. Apparently, the partially oxidized portion of the char was preferentially pyrolyzed on vacuum heating at 370°C.

Assuming, then, that Eq. (2) does represent the behavior of K_2 for pure cellulose, Eqs. (1) and (2) can be combined to permit computation of F_T [5], the ratio of the depolymerization to the char-forming processes at any temperature:

$$F_T = 1.06 \times 10^5 \exp(-12,440/RT) \tag{5}$$

where

$$F_T = (K_B)_T / (K_2)_T \tag{6}$$

Thus, even at 220°C, where weight loss is extremely slow, the depolymerization process accounts for some 25% of the pyrolysis. The competing processes proceed at equal rates at 268°C. Even at 325°C, where pyrolysis is quite rapid, the depolymerization process accounts for only 75% of the reaction, and this fraction reaches 86% at 370°C.

The use of this model to compute $W_{T,\infty}$ [5], the residual weight after infinite time at temperature T, requires a definition of "infinity." Since K_5 has a nonzero value at 370°C, its values at temperatures below 260°C may be negligible, but they cannot be zero. If "infinite time," then, is defined relative to the other reaction rates in the scheme

$$W_{T,\infty} = 0.4136K_2/(K_B + K_2) = 0.4136/(F_T + 1) \tag{7}$$

However, such a residue would continue to lose weight very slowly at temperature T (and, of course, more rapidly if the temperature were then raised). If "infinite time" is defined relative to K_5,

$$W_{T,\infty} = 0.3210/F_T + 1) \tag{8}$$

VI. ARRHENIUS EQUATIONS FOR THE OTHER RATE CONSTANTS

The data are not sufficiently accurate to permit computation of reliable activation energies for the other steps in the pyrolysis sequence. Values for K_1 can be obtained only from the early portions of the curves. Thus, they are highly dependent upon the selection of t_0, the time of start of the isothermal process, and of W_0, the initial sample weight. The error in selecting t_0 is negligible at 226°C, but becomes increasingly important at the higher temperatures. During warmup, a slight error in W_0, important only to the early portion of the curve, is introduced both as a result of zero drift with changing temperature and by loss of any extraneous bound volatiles [3].

An interesting problem arises with K_3 and K_4, illustrating an additional limitation to following the progress of such a sequence of reactions by weight measurements alone. Note that these two constants have similar effects over similar portions of the pyrolysis curve (Table II). In fact, the equations are such that an exact interchange of values for K_3 and K_4, while affecting the time curves for components C and D, produces no change in the curve for component E and consequently, since $W_c = w_d = 1$, no change in the overall weight loss curve. The leeway thus provided in selection of the appropriate values for either K_3 or K_4 (see example at 226°C in Table III) precludes an accurate determination of activation energy.

Finally, K_5 plays too small a role in the pyrolysis of the "pure" samples (and too variable a role in the pyrolysis of the "contaminated" samples) to be considered much more than just an "error term" in these computations.

VII. CONCLUSIONS

When the course of a pyrolysis is followed by measuring weight change as a function of time, it is possible to derive kinetic data for reaction steps in which no weight change occurs. Recognition of this fact has led to the determination of a new set of values for the rate constants and empirical formulas for each step in a previously

proposed pyrolysis scheme. Thus modified, the model elimi-
nates previous inconsistencies and gives theoretical pyrolysis curves consistent with experimental data at other
temperatures. The Arrhenius equation for K_B in this scheme
and the somewhat less reliable Arrhenius equation for K_2
permit computation of the relative contribution of the
depolymerization and the char-forming processes as a function of temperature. For pyrolyses comparable to the 1000
hour experiment, the last step in the char-forming sequence
involves a weight loss corresponding to the formation of
$C_8H_6O_2$ per repeating cellobiose unit in the original cellulose. For longer times, at higher temperatures, or for
cellulose contaminated with inorganic catalysts, an additional step leading to $(C_7H_4O)_n$ occurs. It appears that the
inorganic impurities have little direct effect on K_B but
markedly increase K_2 and, even more markedly, K_5.

ACKNOWLEDGEMENTS

Thanks are due Dr. Ellis F. Darley, University of
California Statewide Air Pollution Research Center (SAPRC),
and Dr. Amos S. Newton, Lawrence Berkeley Laboratory, for
their help in many ways throughout the course of this project.
The work was supported in part by Grant AP00658 from the U.S.
Environmental Protection Agency and GP34494 from the National
Science Foundation to SAPRC. Those receiving such support in
obtaining the experimental data or developing the computer
programs include Drs. Meir Weinstein, Adelina C. Javier-Son,
and Hsiukang Yow, and Mr. Alvin C. Ling and Ms. Yman Teng.

REFERENCES

1. A. Broido and M. Weinstein, in "Thermal Analysis,"
    (Hans. G. Wiedemann, ed.) Vol. 3, pp. 285-296,
    Birkhauer, Basel, 1972.
2. F. J. Kilzer and A. Broido, Pyrodynamics 2, 151-163
    (1965).
3. A. Broido and M. Weinstein, Combustion Sci. Technol. 1
    279-285 (1970).
4. F. A. Williams, private communication.
5. A. Broido and Maxine A. Nelson, Combust. Flame, 24
    263-268 (1975).

FLASH PYROLYSIS OF HOLOCELLULOSE
FROM LOBLOLLY PINE BARK

P. Fang and G. D. McGinnis

Forest Products Utilization Laboratory
Mississippi State University
Mississippi State, Mississippi

I. INTRODUCTION

Disposal of the vast amounts of bark has been one of
the major problems facing the wood industry [1]; however,
this condition has changed due to the energy crisis. Presently, more and more of the bark residue is being used as
fuel in the wood industry. It can replace the more costly
petroleum-based fuel. On the other hand, bark could also be
a potentially renewable source of organic chemicals if an
efficient and economical method of converting bark into commercially useful organic chemicals could be developed.
Pyrolysis of bark has long been studied [2-5] as a possible
method for this conversion. However, little progress has
been made because of the lack of adequate knowledge about the
chemical and thermal properties of bark components.

Loblolly pine bark, one of the major residue sources in southeastern United States, is rich in a variety of chemicals. Studies done [6] in this laboratory showed that it contains 41% of holocellulose with lesser amounts of lignin, phenolic acids and extractives. Earlier thermogravimetric analysis studies [7] indicated that the holocellulose fraction played an important role in the thermal decomposition of whole bark. Although intensive studies [8-10] have been done on pyrolysis of cellulosic materials from wood, very few reports on holocellulose from bark are available. An adequate understanding on thermal behavior of this material would help in the development of utilization of bark through pyrolysis.

In this study, the bark holocellulose was investigated; the pyrolytic products were identified and the effects of temperature and the influence of inorganic additives were determined. The holocellulose samples were pyrolyzed by the method of continuous flash pyrolysis at temperatures from 100° to 600°C at intervals of 50°C. The volatile products were separated and identified in a continuous flow system, using a gas chromatograph/vapor phase thermal dissociation technique.

II. EXPERIMENTS

A. Preparation of Samples

The holocellulose sample was obtained by removing lignin from the extractive-free pine bark by using a sodium chloride-acetic acid method [13]. The yield was 41%, based on the oven-dried weight of extractive-free bark. The holocellulose sample was ground to a fine powder (40 mm diameter). The detailed chemical composition can be found in a previous publication [6]. Samples with additives were prepared by mixing the holocellulose with calculated portions of $ZnCl_2$ or NaOH in a minimum amount of methanol or H_2O to dissolve the additives and assume uniform distribution. The solvent was removed under vacuum at 50°C.

B.   Flash Pyrolysis

Pyrolysis was conducted between 100°C and 600°C at intervals of 50°C by using a Fisher/Victoreer model 27259 pyrolyzer. The temperature was calibrated by thermocouples. Samples of 10 mg were introduced into a platinum boat and pyrolyzed under helium flow at 60 ml/min. The products were swept directly into a Perkin-Elmer 3920 gas chromatograph with a Porapak Q column. Both TC and FID detectors were used. The column temperature was programmed between 70° and 210°C at the rate of 8°C/min. Quantitative determinations were made mainly by gas chromatography using the integrated peak areas for calculation.

C.   Vapor-Phase Thermal Fragmentation

This analysis was done by following the technique developed by Levy and Paul [14]. The individual compounds separated from the GC system were transferred to a gold pyrolysis reactor (T = 650° or 700°C) (Chemical Data System, Inc. model 1200). In the reactor, the compounds were thermally fragmented, and the fragments were characterized through a second gas chromatograph. The pyrolysis products were identified by the comparison of the thermally dissociated fragments with the authentic samples.

III.   RESULTS AND DISCUSSION

Untreated holocellulose was first investigated by flash pyrolysis at the temperature of 100-600°C at intervals of 50°C. The pyrolytic products were swept directly into a gas chromatograph for analysis without trapping. The chromatograms obtained from pyrolysis at 300°C, 350°C and 400°C are shown in Figure 1. Table I provides the list of major identified products that are in agreement with those previously reported on pyrolysis of carbohydrates [8-10]. The temperature effect on the distribution of pyrolytic products is shown in Figure 2.

Pyrolysis at 100°C largely removed the absorbed water in the holocellulose sample. When the temperature was elevated to 150°C, small loss of moisture still could be

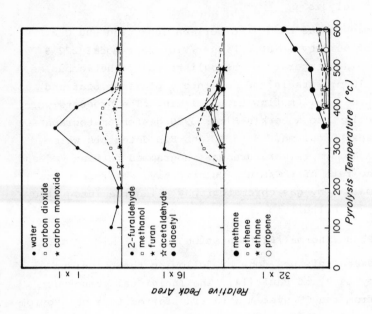

Fig. 2. Effect of temperature on the distillation of volatile products (uncatalyzed).

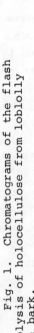

Fig. 1. Chromatograms of the flash pyrolysis of holocellulose from loblolly pine bark.

TABLE I.

Pyrolysis Products of Holocellulose from Loblolly Pine Bark

No.	Compound	Total yield (%)			Identification method[a]
		Neat	5% $ZnCl_2$	5% NaOH	
1-4	Small molecular weight hydrocarbons	0.40	0.43	1.80	a,b
5	Methanol	1.12	1.72	0.65	a,b
6	Acetaldehyde	0.21	0.15	0.34	a,b,c
7	Acrylaldehyde	0.07	0.05	0.09	a,b,c
8	Furan	0.52	0.64	0.47	a,b
9	Acetic Acid	1.39	0.54	1.04	a,b
10	Diacetyl	0.23	0.24	0.31	a,b,c
11	1-Hydroxy-2-propanone	0.06	0.02	0.22	a,b,c
12	2-Furaldehyde	0.51	0.77	0.06	a,b,c
	Carbon monoxide	5.28	5.80	8.98	b
	Carbon dioxide	11.04	11.81	20.42	b
	Water	37.25	42.51	37.22	b
	Char	20.17	34.83	28.04	--

a. Identification method: (a) Comparison of retention time and addition of known compound; (b) vapor phase fragmentation; and (c) formation of a 2,4-dinitrophenylhydrazone derivative.

detected. In order to determine the beginning of the dehydration reaction, pyrolysis was done between 100-180°C at intervals of 10°C. It was found that dehydration accompanied by decarboxylation started at 150-160°C. Only minor traces of other products were detected at 150°C and 200°C. At 250°C, small amounts of methanol, furan, acetic acid, 2-furaldehyde, and carbon monoxide were observed. The evolution of H_2O, CO, and CO_2 started in a vigorous rate when pyrolysis was conducted at 300°C. The maximum yields of H_2O and CO_2 were observed at 350°C, and at 400°C for CO. In addition, methanol and 2-furaldehyde were also obtained at

their highest yield in the temperature region of 300-350°C.
The production of other major products, such as acetaldehyde, acidic acid, diacetyl and 1-hydroxy-2-propanone
rapidly increased at 350°C, and the maximum yield occurred at
400°C. Above 400°C, the whole production of volatile
products decreased, except for the low molecular weight hydrocarbons in which methane predominated. The final nonvolatile residue was 20%, based on the starting holocellulose
sample.

A previous study [10] found that a number of monosaccharides, disaccharides, and cellulose all gave the same
volatile products on pyrolysis at 300-500°C for a short time.
However, there is a considerable amount of evidence that the
distribution of the pyrolytic products upon temperatures is
related to the structure and the crystallinity of the polysaccharides. Despite the controversy about the mechanism of
thermal degradation of carbohydrates, several studies [11]
have shown that pyrolysis of cellulose begins at about 300°C
and proceeds very rapidly at 350°C. Hemicellulose begins to
pyrolyze at a lower temperature. In this study, the large
amount of methanol and 2-furaldehyde found in pyrolysis at
300°C was mainly contributed by the bark hemicellulose
fractions.

The higher proportion of 2-furaldehyde to other major
carbonyl products found at 300°C came from the D-xylose
units, which are present in the xylans fraction of bark. In
general, pentoses give much higher yield of 2-furaldehyde,
whereas hexoses give practically no 2-furaldehyde and more
lower molecular weight carbonyl compounds.

The effects on pyrolysis of introducing inorganic
additives are shown in Figure 3 and Table I. Earlier studies have shown that $ZnCl_2$ and NaOH modified the thermal
degradation of cellulosic materials [10] and bark phenolic
acid [12]. These catalysts were mixed with the holocellulose
in the ratio of 5-20% for NaOH and 5-30% for $ZnCl_2$. The
pyrolysis was done under the same conditions as the uncatalyzed pyrolysis. The addition of 5% $ZnCl_2$ shifted the
pyrolytic decomposition. The $ZnCl_2$ lowered the temperature
by about 50°C and generated more furan compounds.

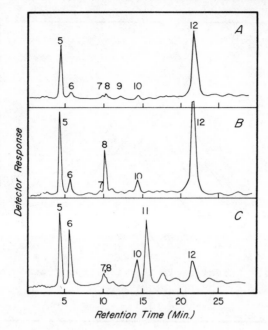

Fig. 3.   Effect of inorganic compounds
on pyrolysis of holocellulose at 200°C;
(A) no treatment, (B) 5% ZnCl$_2$, (C) 5% NaOH.

Dehydration and decarboxylation drastically increased at
250°C.   The catalyst effect on the productions of 2-furalde-
hyde and methanol versus temperature is shown in Figure 4
and 5.   Compared to uncatalyzed holocellulose, the total
yield of these two products from a 5% ZnCl$_2$ sample was
increased to about 50% and 55%, respectively.

Increasing the concentration of ZnCl$_2$ about 5% did not
cause a further increase in the amount of 2-furaldehyde but
yielded more methanol.   Increased amounts of ZnCl$_2$ also
cause a lowering of the decomposition temperature.   Samples
containing 20% and 30% ZnCl$_2$ started considerable decomposition around 200°C.   The production of carbonyl compounds
followed the same temperature pattern, but the yield was
lower than for untreated holocellulose.   The nonvolatile
residue was found to be 32-34% based on holocellulose.

The addition of NaOH also shifted the decomposition to
lower temperatures. The distribution of products was quite

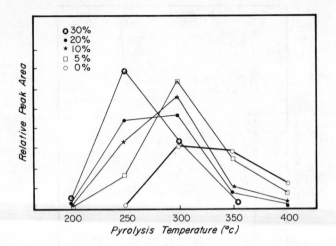

Fig. 4.   Effect of ZnCl$_2$ on the production of
2-furaldehyde.

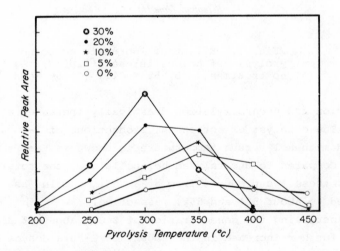

Fig. 5.   Effect of ZnCl$_2$ on the production of
methanol.

different from the case of ZnCl$_2$. Pyrolysis of 5% NaOH sample at 250°C evolved more H$_2$O and CO$_2$ than the uncatalyzed sample. However, only a small amount of the other volatile compound could be observed. The evolution of CO$_2$ was even more vigorous than in ZnCl$_2$, but less water evolved. When the pyrolysis temperature was elevated to 300°C, much more acyclic carbonyl compounds were formed. The catalytic effects of NaOH on acetaldehyde and diacetyl are shown in Figures 6 and 7. As shown in these two figures, increasing the concentration of NaOH did not change the decomposition pattern very much. The formation of 2-furaldehyde was depressed through the whole pyrolysis process. The most drastic change was the high yield of low molecular weight hydrocarbons as shown in Figure 8. Methane was found to increase by about 3.7 to 1; ethene, 3 to 1; and ethane, 10 to 1 when compared to the untreated holocellulose. The total yield of nonvolatile material was 25-28% based on holocellulose.

By examining the flash pyrolysis pattern of bark holocellulose, the relationship between the volatile products, the temperature, and the concentration of inorganic additives has been determined. This basic investigation will, we hope, help the development of pyrolytic methods for utilizing bark material.

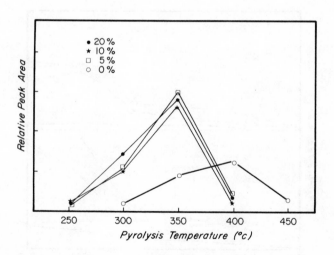

Fig. 6. Effect of NaOH on the production of acetaldehyde.

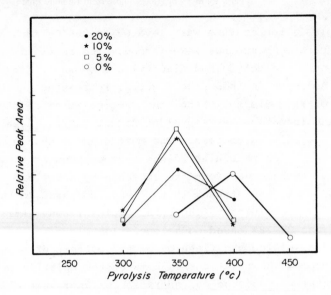

Fig. 7.   Effect of NaOH on the production of diacetyl.

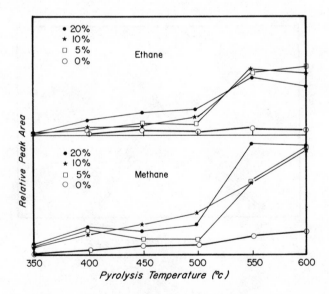

Fig. 8.   Effect of NaOH on the production of ethane and methane.

REFERENCES

1. Wood Products Sub-Council, <u>Forest Prod. J</u>. <u>21</u> (9), 33
    (1971).

2. H. N. Euler, H. Hasselquist, U. Loov, and S. Edelo,
    <u>Festskr. J. Arvid Hedvall</u> p. 169 (1948).

3. E. F. Kurth and C. V. S. Ratnam, <u>TAPPI</u> <u>33</u>, 517, (1950).

4. A. H. Vroom, <u>Pulp Paper Mag. Can</u>. <u>53</u>, 121, (1952)

5. S. Z. Chow and K. J. Pickles, <u>Wood Fiber</u> <u>4</u> (3), 166
    (1971).

6. G. D. McGinnis and S. Parikh, <u>Wood Science</u> <u>7</u>, 295,
    (1975).

7. P. Fang and G. D. McGinnis, <u>Wood Fiber</u> (in press).

8. R. F. Schwenker and L. R. Beck, <u>J. Polym. Sci</u>. Part C <u>2</u>,
    331 (1963).

9. F. Shafizadeh, <u>Advan. Carbohyd. Chem</u>. <u>23</u>, 419 (1968).

10. K. Heyns and M. Klier, <u>Carbohyd. Res</u>. <u>6</u>, 436 (1963).

11. F. Shafizadeh and G. D. McGinnis, <u>Carbohyd. Res</u>. <u>16</u>, 273
    (1971).

12. P. Fang and G. D. McGinnis, unpublished result.

13. B. L. Browning, "Wood Chemistry," Vol. 2, p. 1138.
    Reinhold, New York, 1952.

14. E. J. Levy and D. G. Paul, <u>J. Gas Chromatog</u>. <u>5</u>, 136,
    (1967).

COMPARISON OF THE THERMAL DEGRADATION PRODUCTS

OF α-CELLULOSE AND DOUGLAS FIR UNDER

INERT AND OXIDATIVE ENVIRONMENTS

F. D. Hileman, L. H. Wojcik,
J. H. Futrell, and I. N. Einhorn

Flammability Research Center
Department of Chemistry
University of Utah
Salt Lake City, Utah

I. INTRODUCTION

Considerable interest has been shown in recent years in the area of toxic products formed during fire situations. The primary interest of our research group is to identify the thermal decomposition products of both natural and synthetic materials. Within this scope, the work presented in this paper was to determine the decomposition products of Douglas fir under varying conditions. When it was necessary, new analytical techniques were developed to aid in the monitoring or the identification of these products. A parallel study using α-cellulose was also carried out, since the cellulose decomposition products would be a subset of the wood decomposition products. This would simplify the identification of products by allowing them to be classified as

resulting primarily from cellulose or from the lignins,
resins, tannins, and waxes, which are also present in wood
[1].

II. EXPERIMENTS

A. Thermal Analysis

In these studies, all thermal analyses were carried out
on a Mettler Thermal Analyzer model 1. Provision has been
made for the trapping of the effluents from the oven for subsequent analysis of the products. This analysis was performed using a Hewlett-Packard model 7620A gas chromatograph
coupled to a Hewlett-Packard model 5930A mass spectrometer
complete with data system [2]. The effluent gas from the
Mettler oven was also repeatedly sampled with a Gow Mac model
560 gas chromatograph. This chromatograph is equipped with a
10 inch molecular sieve 5A column and a 2 foot Porapak Q
column, which permits rapid (2 min) repetitive light-gas
(O_2, N_2, CO, co_2, H_2O) analysis.

B. Elemental Analysis

Elemental analysis of samples was performed on a Perkin-
Elmer model 240 Elemental Analyzer. Samples of materials
were analyzed both prior to thermal analysis and after various stages of thermal decomposition.

C. Calibration of Pyroprobe

A Chemical Data Systems Pyroprobe model 120 was used
when fast heating rates (200°C/sec) were desired. The pyroprobe and quartz sample holder were calibrated for true
pyrolysis temperatures (Fig. 1) using a chromel-alumel
thermocouple fabricated from 40 gauge wire. The thermocouple
output was fed to a Oscillomink P oscillographic recorder
with response time of 100 Hz/sec. A comparison of the heating rates with and without sample is shown for 1.5 mg of
Douglas fir in Figure 2. The plateau in the heating curve
with sample indicates the temperature where the power needed
to sustain the pyrolysis process equals the power being

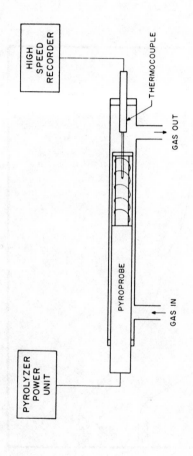

Fig. 1. Apparatus for the measurement of true pyrolysis temperature.

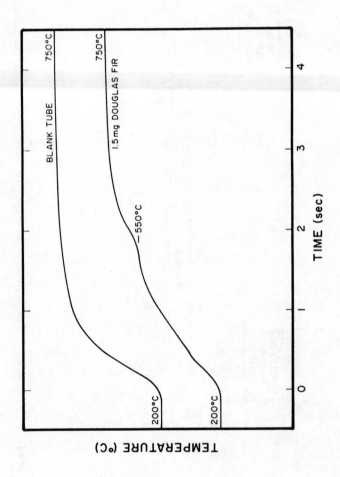

Fig. 2. Measurement of the pyrolysis temperature of Douglas fir.

applied.   This temperature has been defined by Levy as the
true pyrolysis temperature [3].   The true pyrolysis temperature was determined for all samples in all environments used
in this study.

D.   Direct Pyrolysis-Chemical Ionization Mass Spectrometry

        Figure 3 depicts the experimental equipment necessary
for direct pyrolysis-chemical ionization mass spectrometry
(DP-CIMS).   The pyroprobe with 1 to 2 mg of sample was sealed
into a heated all-glass 10 ml expansion bulb.   This bulb was
connected via 1/16 inch glass-lined stainless steel tubing to
the ion source of a computerized Varian MAT CH-7 mass

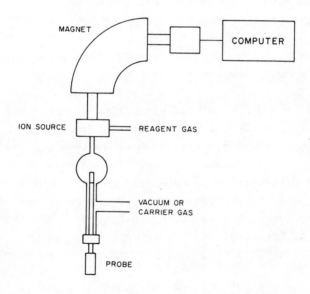

        Fig. 3.   Configuration for the direct pyrolysis-mass
spectrometry of materials.

spectrometer modified for chemical ionization [4]. The
expansion bulb was then filled with argon to establish a
10 ml/min sweep gas flow rate. Methane was then added to the
argon in a ratio of 1 to 5 at the ion source to serve as
chemical ionization reagent gas [5]. Background spectra were
taken at 10 sec intervals, with the data being stored by the
computer. The pyroprobe was subsequently fired and the
products swept into the ion source by the argon. Repeated
mass spectra were obtained at 10 sec intervals until the
product supply had been depleted. To avoid interfacing low-
mass reagent ions, mass spectra were recorded from m/e 50 to
m/e 800. These spectra were stored as mass versus absolute
peak intensities, and they could be averaged, substracted,
and manipulated in any desired manner.

E.   Gas Chromatograph-Mass Spectrometry

The gas chromatograph-mass spectrometry (GC-MS) analysis of thermal decomposition products involved pyrolyzing
1 to 2 mg samples directly onto the chromatographic column
using the calibrated pyroprobe. The gas chromatograph-mass
spectrometer system was the same system described in the
thermal analysis section. If a pyrolyzing atmosphere other
than the gas chromatograph carrier gas was desired, then a
simple trapping device was used that permitted cryogenic
collection of the volatiles with subsequent heat desorption
onto the chromatographic column. Either Porapak Q or Tenax
was used as the trapping material, depending on whether low
or high molecular species were to be analyzed. Four chromatographic substrates were used in this work--Chromosorb 101,
Porapak Q, 10% UC-98 on Chromosorb W, and 10% Carbowax 20M on
Chromosorb W. All columns were 8 ft x 1/8 in O.D. stainless
steel.

Identification of eluted compounds was based on comparison of their mass spectra to a reference library [6] and by
comparison of their retention indexes to either a reference
library [7] or to estimated values calculated from this
library [8].

F.   Extraction of Lignin

     Attempts were made to separate the lignin from the
Douglas fir with as little change to the lignin as possible.
The extraction procedure described by Bolker et al. [9] was
followed, which involved the soxhlet extraction of 3 to 4 gm
wood sawdust with dry acidified (HCl) dimethoxypropane in
dioxane. Purification was achieved by neutralization of the
extracting solution with sodium bicarbonate followed by
reprecipitation of the noncellulosic materials in the
pentane. A second method described by Bolker used acidified
(HCl) aqueous dioxane as the extracting solution followed by
the same neutralization/purification procedure. Yields of
the lignin in either extraction ranged from 20 to 30% of the
original weight of the wood sawdust.

G.   Materials

     The Douglas fir used in this work was provided by
Weyerhauser Company, Longview, Washington. The wood sample
was grated to form a coarse sawdust. α -cellulose was
obtained from the National Bureau of Standards in the form of
a loosely compressed sheet. All samples were vacuum dried
overnight at 55°C to remove any residual water and, in the
case of lignin, to remove any absorbed extracting or purify-
ing solvent.

III.   RESULTS AND DISCUSSION

A.   Thermal Analysis

     The thermogram of cellulose in helium (Fig. 4) indi-
cates that pyrolysis begins at 320°C and has a maximum rate
of weight loss at 375°C with little remaining char. When the
cellulose was heated in air, the decomposition again begins
at 320°C, but with a maximum rate of weight loss (Fig. 5) now
shifted down to 350°C. The endothermic reaction shown in
helium was replaced by a strongly exothermic decomposition in
air. At 360°C, a char consisting of 20% by weight of the
original sample remained and slowly eroded with increasing

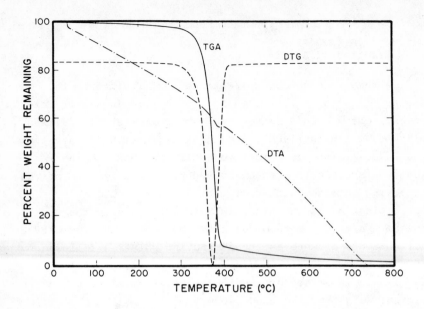

Fig. 4.   Thermal analysis of α-cellulose in helium heated at a rate of 10°C/min.

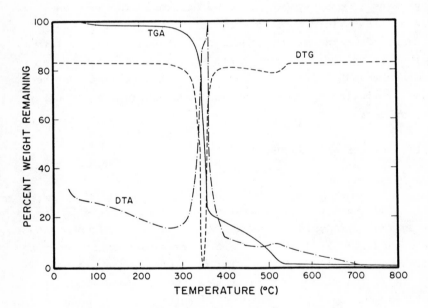

Fig. 5.   Thermal analysis of α-cellulose in air heated at a rate of 10°C/min.

temperature. The thermal analysis for Douglas fir in helium
(Fig. 6) indicates sample decomposition beginning around
250°C, considerably lower than cellulose, with a slightly
endothermic reaction. The maximum rate of weight loss occurs
at 370°C, similar to cellulose, with approxinately 20% char
formation. When Douglas fir was heated in air (Fig. 7), an
exothermic decomposition begins again at 250°C. The maximum
rate of weight loss has shifted down to 335°C, 15°C lower
than for cellulose in air. At 360°C, a stable char has formed
that had an elemental composition of 74.18% C, 3.34% H,
22.52% O (by difference). This char again slowly eroded in
air with increasing temperature.

B.   Decomposition Product Analysis

The direct pyrolysis-chemical ionization mass spectrometry analysis (DP-CIMS) of α-cellulose and Douglas fir
provides a unique method for initially examining the decomposition products of these materials.   In the methane chemical ionization process, the energy transfer to the sample
molecule during protonation is normally low enough that
insufficient energy is available for molecule fragmentation.
Thus, the various ions observed in DP-CIMS of Douglas fir
and cellulose (Fig. 8) result from the fragmentation by the
400°C pyrolysis and not from any fragmentation during
ionization. A particular advantage of this technique was the
lack of fragmentation, implying a strong protonated molecular
ion (M + 1), and also a strong (M + 2) peak resulting from
the natural abundance of carbon-13. The M + 1 peak allows a
determination of the molecular weight of the species in
question, and the M + 2 peak allows the determination of the
elemental abundance. As an example, the m/e 127 peak in the
α-cellulose spectra was calculated to have an elemental
composition of $C_6H_6O_3$, which was determined to be 5-hydroxy-
methyl-2-furaldehyde from gas chromatographic analysis.

The thermal decomposition products of cellulose form a
subset of the decomposition products from Douglas fir [10].
Thus, a synthetic generation of the mass spectra of the
thermal decomposition products from noncellulose components

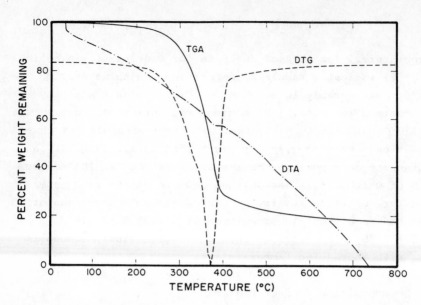

Fig. 6. Thermal analysis of Douglas fir in helium heated at a rate of 10°C/min.

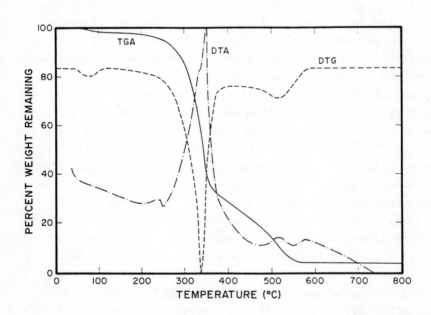

Fig. 7. Thermal analysis of Douglas fir in air heated at a rate of 10°C/min.

58

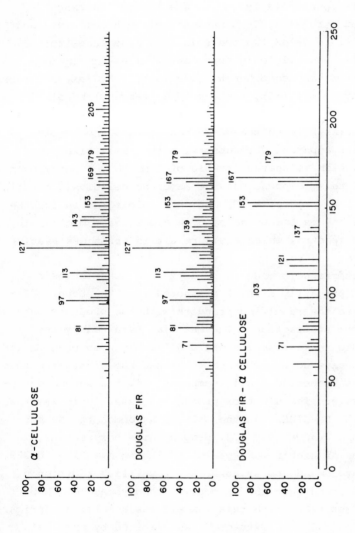

Fig. 8. Direct pyrolysis of Douglas Fir and ∝-Cellulose and the difference in their products.

in wood can be affected by subtracting from a DP-CIMS mass
spectrum of Douglas fir the DP-CIMS mass spectra of a pro-
portionate amount of cellulose. Figure 8 depicts a subtracted spectra for a 400°C argon atmosphere pyrolysis. The
subtracted spectrum reveals the presence of many noncellulose
components, particularly in the m/e 130 to 180 range.
Subsequent pyrolysis GC-MS analysis of both cellulose and
Douglas fir revealed the peaks in this mass range to be
essentially noncellulosic components. These components
include p-methoxyacetophenone (m/e 151), 4-hydroxy-3-methoxy-
benzaldehyde (m/e 153), and 2-methoxy-4-propylphenol
(m/e 167), as detailed in the GC-MS product list in Table I.
Discrepancies between the GC-MS results and the DP-CIMS
results are apparent when observing the small size of
m/e 139, which is characteristic of 2-methoxy-4-methylphenol.
At 400°C, this compound was the major noncellulosic material
decomposition product by GC-MS. A possible reason for the
inconsistency is the presence of a significant m/e 139 in the
cellulose spectra, which could result in erroneous results in
the subtraction routines.

Attempts were made to pyrolyze the noncellulosic
material (primarily lignin) extracted from the Douglas fir
and compare the resulting products with the products obtained
by spectral subtraction. The material obtained by the
aqueous dioxane extraction was analyzed by both DP-CIMS and
GC-MS and gave no resemblance to either the subtracted spectra or the suspected lignin components in the chromatographic analysis. The dimethoxypropane-dioxane lignin was then
examined by DP-CIMS, and the only peak common to the subtracted spectra was m/e 179, propenyldimethoxybenzene. A
succession of higher mass peaks, including m/e 219 and 259,
were observed and appear to be the repetitive addition of
propenyl groups onto the propenyldimethoxybenzene. When the
wood sawdust from which this noncellulosic material (primarily lignin) had been extracted was examined by pyrolysis
GC-MS, the components attributed to noncellulosic materials
(primarily lignin) were still present but significantly
reduced. Thus, the difference in the subtracted spectra and
the spectra of the extracted material could result from the

TABLE I

Douglas Fir--Inert Atmosphere

Product	Yield (%)				
	340°C	370°C	400°C	450°C	550°C
Carbon monoxide	0.7	1.1	2.8	5.2	21.2
Methane	--	Trace	0.2	0.9	2.0
Carbon dioxide	2.1	3.2	5.7	5.6	5.7
Ethene	--	--	Trace	Trace	1.0
Ethane	--	--	Trace	Trace	0.3
Water	10.3	13.5	17.9	14.3	15.3
Propene	--	--	0.1	0.2	1.6
Methanal	0.2	Trace	0.1	0.4	0.9
Methanol	0.2	0.2	0.3	0.3	0.4
Ethanal	Trace	0.1	0.3	0.5	1.4
2-Methylpropene	--	--	--	Trace	0.3
Propenal	0.1	0.2	0.5	0.5	0.7
Furan	--	--	0.1	0.1	0.2
2-Oxopropanal	0.2	0.4	0.7	0.8	0.5
Hydroxyethanal	Trace	0.5	0.7	0.6	0.7
Ethanoic acid	Trace	0.5	0.6	0.5	0.7
2,3-Butanedione	--	Trace	0.2	0.4	0.3
2-Hydroxypropanone	--	0.2	0.8	0.7	0.5
Furfural	0.1	0.5	0.5	0.5	0.4
2,3-Pentanedione	--	Trace	0.1	0.1	0.1
5-Methylfurfural	--	Trace	0.1	0.1	--
Furfuryl alcohol	--	Trace	0.1	0.1	0.2
Cyclohexanone	--	Trace	Trace	Trace	--
2-Methyl-2-butenal	--	--	--	--	0.1
2-Hydroxy-3-methyl-2-cyclopenten-1-one	--	0.1	0.1	0.1	Trace
o-Methoxyphenol	--	0.1	0.3	0.3	0.2
NI	--	0.1	0.1	0.1	Trace
2-Methoxy-4-methylphenol	Trace	0.1	0.8	0.7	0.4
NI	--	Trace	0.1	0.1	0.2
2-Methoxy-4-methylanisole	--	0.1	0.3	0.2	0.1
4-Oxophentanoic acid	--	0.2	0.1	0.2	0.4
NI	--	0.1	0.1	0.1	Trace
4-Hydroxypentanoic acid	--	--	0.2	0.2	0.2
p-Methoxyacetophenone	Trace	0.2	0.4	0.3	0.2
NI	--	0.2	0.3	0.2	0.2
2-Methoxy-4-propenylphenol	Trace	0.2	0.3	0.2	0.1
5-Hydroxymethyl-2-furaldehyde	Trace	0.3	0.3	0.2	0.2
4-Hydroxy-3-methoxybenzaldehyde	--	Trace	0.1	0.1	--
2-Methoxy-4-propylphenol	___	Trace	0.2	0.1	--

change in noncellulosic material during extraction or possibly some gas phase or solid phase reaction between cellulose
and noncellulosic material decomposition products.

The pyrolysis products of Douglas fir at various temperatures determined by gas chromatography are given in
Table I as the percentage of product formed from the original
sample weight. Sample sizes were typically 1 to 2 mg with a
10 sec pyrolysis time. At low pyrolysis temperatures, the
principal components that were formed were carbon dioxide and
water. Since all samples were carefully dried before pyrolysis, the water formed results from the dehydration of the
cellulose in wood [11, 12]. A fresh sample of Douglas fir
was then pyrolyzed at 370°C, at which point the decomposition
products increase significantly. These products, 2-oxo-
propanal, hydroxyethanal, ethanoic acid, 2-hydroxypropanone,
furfural, and 5-hydroxymethyl-2-furaldehyde can all be
related to the decomposition of the cellulosic portion of
wood, since all of these products also result from the
pyrolysis of cellulose. Examining the TGA of cellulose in
helium reveals the 370°C temperature to be approximately the
region of maximum weight loss. As the temperature of pyrolysis was increased to 400°C, the aromatic compounds typical of
the noncellulosic portion of wood make a dramatic appearance.
This was particularly true for 2-methoxy-4-methylphenol, the
main noncellulosic material decomposition product from this
Douglas fir. The cellulosic decomposition products increase
in concentration slightly, with many of them reaching their
maximum levels at this temperature. Carbon dioxide and
carbon monoxide levels have been continually increasing with
increasing temperatures up to 400°C. As the pyrolysis
temperature was increased to 450°C and then 550°C, the production of carbon dioxide remains essentially constant,
while the carbon monoxide continues to increase to very high
levels of production (21.2%). The decomposition of cellulose gave an identical pattern, with carbon dioxide reaching
a maximum level of production at 400°C. This tends to
indicate that a certain limited amount (5.7%) of decarboxylation can occur in the inert-atmosphere decomposition of
Douglas fir or cellulose. Maximum levels of production of

moderate molecular weight (40-100) decomposition products from either cellulose or noncellulosic material occur at mid range temperatures (400°C). At lower temperatures, the energy is not available for their production, at higher temperatures, they are possibly formed and repyrolyzed to small molecules (carbon monoxide, methane, ethene, propene) before they are able to leave the heated zone.

The pyrolysis products of Douglas fir were also studied by the sequential heating of a sample to progressively higher temperatures. A sample was weighed, then pyrolyzed at 340°C with volatile product analysis, and then reweighed. This same sample was then repyrolyzed at 400°C with product analysis, reweighed, and finally repyrolyzed at 550°C. The distribution of products obtained from this mode of pyrolysis is given in Table II. At moderate temperatures (400°C), the majority of compounds are formed and cover the range from methane to 5-hydroxymethyl-2-furaldehyde. At 400°C, the product distribution obtained by this mode of analysis was, within experimental error, identical to the distribution of products trapped in the 300°C to 400°C range from the Mettler Thermal Analyzer using a heating rate of 45°C/min. At the high pyrolysis temperature (550°C), the product distribution shifts completely to the low molecular weight species.

When Douglas fir was thermally decomposed in the air using high heating rates, the percent residues at 340°C and 370°C (Table III), are lower in air than in inert atmosphere. This would be expected from comparing the TGA curves in Figures 6 and 7. However, at 400°C and 450°C, the char residues were actually greater from oxidative degradation than pyrolysis. When the temperature was increased to 550°C, the oxidative char became less than the pyrolysis char. This was not the situation for $\propto$-cellulose, where the oxidative char was always less than the char formed in an inert atmosphere. Other workers have reported an increase in char yield of cellulose that had been preheated for long periods of time [12]. Since this phenomenon was observed for wood and not for cellulose, it indicates that possibly the lignin was responsible for this char stabilization.

TABLE II

Repeated Pyrolysis of Douglas Fir in an Inert Atmosphere

Product	Yield (%)		
	340°C	400°C	550°C
CO	0.7	2.0	4.2
CH_4	--	0.3	1.3
CO_2	2.1	3.3	1.1
Ethene	--	Trace	0.1
Ethane	--	Trace	0.1
Water	10.3	5.0	2.1
Methanal	0.2	0.2	Trace
Propene	--	--	Trace
Methanol	0.2	0.2	0.1
Ethanal	Trace	0.2	0.1
Propenal	0.1	0.2	Trace
Furan	--	0.4	Trace
2-Oxopropanal	0.2	--	--
Hydroxyethanal	Trace	0.5	--
Ethanoic acid	Trace	0.4	Trace
2,3-Butanedione	--	0.1	--
2-Hydroxypropanone	--	0.5	--
Furfural	0.1	0.2	--
2,3-Pentanedione	--	0.1	--
Furfuryl alcohol	--	0.1	--
o-Methoxyphenol	--	0.3	--
2-Methoxy-4-methylphenol	--	0.8	--
4-Oxapentanoic acid	--	--	Trace
4-Hydroxypentanoic acid	--	Trace	--
p-Methoxyacetophenone	Trace	0.2	--
2-Methoxy-4-propenylphenol	Trace	0.1	--
5-Hydroxymethyl-2-furaldehyde	Trace	0.1	--

TABLE III

Percent Residues of Cellulose and Douglas Fir Decomposed at Various Temperatures in Inert and Oxidative Environments

Temp.	Cellulose		Douglas fir	
°C	Inert	Oxidative	Inert	Oxidative
320	--	$92.8^{\pm}3.7$	--	--
340	$93.6^{\pm}2.6$	$74.2^{\pm}3.4$	$72.3^{\pm}6.3$	$62.6^{\pm}3.9$
370	$66.7^{\pm}5.8$	$39.4^{\pm}1.5$	$51.4^{\pm}2.2$	$46.1^{\pm}5.6$
400	$32.7^{\pm}6.3$	$22.2^{\pm}5.0$	$36.1^{\pm}4.0$	$40.5^{\pm}2.4$
450	$20.2^{\pm}4.5$	$16.6^{\pm}0.8$	$26.7^{\pm}4.2$	$29.3^{\pm}4.6$
550	$9.7^{\pm}2.2$	$5.7^{\pm}2.2$	$21.9^{\pm}4.0$	$18.8^{\pm}3.7$

Table IV presents a comparison of the inert and oxidative degradation products for Douglas fir heated to 400°C at 200°C/sec. With the exception of carbon dioxide, the product distributions are essentially identical. This indicates that under fast heating rates, the oxygen does not have sufficient time to enter into the degradation process. Because of the necessity of trapping the oxidative degradation components, carbon monoxide and methane could not be analyzed.

In the sequential heating of a Douglas fir sample in air, the product distribution (Table V) appears quite different from a similar analysis under inert conditions. In repetitive pyrolysis, the CO_2 concentration was first observed to increase and then decrease with increasing temperature. With oxidative degradation, the CO_2 continually increases with increasing temperature. This indicates that the oxygen is now able to carry out an oxidation of the previously heated surface. Another significant point is that the concentration of many of the higher molecular weight species, especially the 2-methoxy-4-methylphenol has been reduced. This could be due to either oxidation of these compounds or the possibility of the lignin stabilized char being formed as previously hypothesized.

TABLE IV

Inert and Oxidative Degradation Products of Douglas Fir
at 400°C Using a Fast Heating Rate

Product	Yield (%)	
	Inert	Oxidative
Carbon dioxide	5.7	11.8
Ethene	Trace	--
Ethane	Trace	--
Water	17.9	14.2
Propene	0.1	Trace
Methanal	0.1	0.1
Methanol	0.3	0.3
Ethanal	0.3	0.3
Propenal	0.5	0.3
Furan	0.1	Trace
2-Oxopropanal	0.7	0.6
Hydroxyethanal	0.7	0.9
Ethanoic acid	0.6	0.8
2,3-Butanedione	0.2	--
2-Hydroxypropanone	0.8	0.4
Furfural	0.5	0.5
Methylfurfural	0.1	0.1
Furfuryl alcohol	0.1	0.1
O-Methoxyphenol	0.3	0.3
2-Methoxy-4-methylphenol	0.8	0.9
2-Methoxy-4-methylanisole	0.3	0.2
4-Oxopentanoic acid	0.1	0.1
4-Hydroxypentanoic acid	0.2	0.2
p-Methoxyacetophenone	0.4	0.4
2-Methoxy-4-propenylphenol	0.3	0.3
5-Hydroxymethyl-2-furaldehyde	0.3	0.3

TABLE V

Douglas Fir in an Oxidative Atmosphere

Product	Yield (%)		
	340°C	400°C	550°C
Carbon dioxide	3.7	7.6	21.5
Ethene	--	Trace	Trace
Ethane	--	Trace	Trace
Water	10.4	7.5	7.9
Methanal	0.2	0.1	--
Methanol	0.1	0.3	0.1
Ethanal	Trace	0.1	--
Propenal	Trace	0.1	--
2-Oxopropanal	0.3	0.1	--
Hydroxyethanal	Trace	--	--
Ethanoic acid	0.2	0.1	--
2-Hydroxypropanone	Trace	Trace	--
Furfuals	0.3	0.1	--
Furfuryl alcohol	Trace	Trace	--
o-Methoxyphenol	Trace	0.2	--
2-Methoxy-4-methylphenol	0.1	0.2	--
2-Methoxy-4-methylanisole	Trace	0.1	--
4-Oxopentanoic acid	--	--	Trace
4-Hydroxypentanoic acid	0.1	Trace	--
p-Methoxyacetophenone	0.1	Trace	--
2-Methoxy-4-propenylphenol	0.1	--	--
5-Hydroxymethyl-2-furaldehyde	Trace	--	--

When these fast heating rate results are compared to the slow heating rate oxidation products produced by the Mettler thermoanalyzer, the agreement is not as good as with the inert atmosphere conditions. The products trapped from 250°C to 350°C at 45°C/min tend to be a mixture of the product ratios from both the 340°C and 400°C degradation.

In order to complete the analysis and comparison of inert versus oxidative degradation products, the effluent gas

coming from the Mettler oven was sampled every 2 min during
the periods of significant weight change. The gases of interest were carbon monoxide and carbon dioxide, and when using
an oxidizing atmosphere, oxygen was also monitored. Figure 9
shows the variation of gases concentration and sample weight
with temperature for 65 mg of Douglas fir in helium. Using
the rate of weight loss and the dilution of the sample as
it was swept out of the chamber, the maximum rate of production of CO is approximately 2%, while the maximum rate for
CO_2 is approximately 4% at 380°C. These values agree reason-
ably well for the conversion rates given for the fast heating
of Douglas fir listed for either 370°C or 400°C in Table I.
The evolved gas analysis for 65 mg of Douglas fir heated in
air is given in Figures 10 and 11. Carbon monoxide displays
a very similar elution profile, as exhibited under inert
conditions, and shows approximately the same maximum rate of
production and, more important, the same total amount of
carbon monoxide. This strongly suggests that the carbon
monoxide is produced by the decomposition of the wood and not
from incomplete oxidation. The situation for CO_2 is quite
different. Carbon dioxide levels are significantly higher in
the oxidative degradation than in the inert atmosphere pyrolysis. This production of carbon dioxide results from the
oxidation of the wood or the volatile products that are
formed by the depletion of oxygen as shown in Figure 11.

IV. CONCLUSION

    The higher char formation for cellulose heated in air
compared to helium suggests a possible polymerization process
which could be caused by the presence of oxygen. However,
the lower decomposition temperatures of both α-cellulose
and Douglas fir in air over helium suggest that oxygen is
possibly preventing a polymerization process through the
formation of volatile oxygenated species. In an attempt to
resolve this problem, a sample of cellulose will be heated
to 350°C in air and then rapidly cooled. The sample will
then be reheated in inert atmosphere to see which TGA curve
is followed and if a stabilized char has been formed.

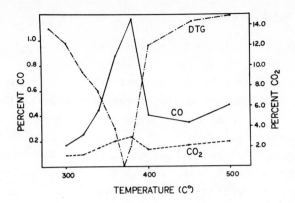

Fig. 9. Effluent gas analysis of Douglas fir in helium.

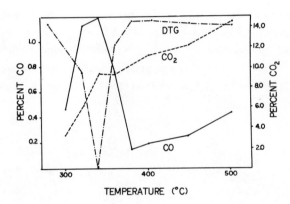

Fig. 10. Effluent gas analysis of Douglas fir in air.

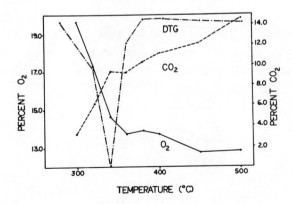

Fig. 11. Effluent gas analysis of Douglas fir in air.

Application of DP-CIMS and the subtraction of spectra has provided a criterion for determining when lignin has been extracted in a form similar to that in wood. The pyrolysis of lignin extracted by several techniques has not yielded a lignin spectrum comparable to that obtained by pyrolyzing wood. This could result from changes in the lignin upon extraction or from the lack of cellulose during the pyrolysis of the lignin. Thus, future studies will involve the mixing of extracting lignin and cellulose to determine if solid or gas phase reactions are occurring.

Finally, the following of the light gas decomposition products of Douglas fir in air and helium has provided new insight into the kinetics of decomposition. Faster and higher-resolution gas chromatographic techniques are being developed to follow an even greater number of products, many of which, due to common molecular weight or serious electron impact fragmentation cannot be followed by a mass spectrometer [12].

ACKNOWLEDGEMENTS

This research was supported by the National Science Foundation and the National Fire Prevention and Control Administration Grant number GI-33650 under its RANN (Research Applied to National Needs) program. The authors are especially indebted to the late Dr. Ralph Long for his many years of continued support of this work. The chemical ionization study was supported in part by the National Science Foundation Grant GP-38125X1.

REFERENCES

1. A. W. Goos, The thermal decomposition of wood, in "Wood
    Chemistry" (L. E. Wise and E. C. Jahn, eds.) (2nd ed.),
    Vol. 2, p. 826. Reinhold, New York, 1952.
2. I. N. Einhorn, D. A. Chatfield, K. J. Voorhees, F. D.
    Hileman, R. W. Mickelson, S. C. Isreal, J. H. Futrell,
    and P. W. Ryan, Fire Research 1 (1976). (Accepted for
    publication.)
3. R. L. Levy, D. L. Fanter, and C. J. Wolf, Analytical
    Chemistry 44, 38 (1972).
4. M. L. Vestal, T. A. Elwood, L. H. Wojcik, and J. H.
    Futrell, Proc. 20th Ann. Conf. Mass Spectrom. Allied
    Topics 20, ASMS, Dallas, Texas, (1971).
5. M.S.B. Munson, Analytical Chemistry 43, 29A (1971)
6. A. Cornu and R. Massot, "Compilation of Mass Spectral
    Data." Heyden and Son, London (1964).
7. "Gas Chromatographic Data Compilation" AMD, 25A S1.
    American Society for Testing Material, Philadelphia,
    (1971).
8. E. Kovats, Advan. Chromatog. 1, 229 (1965).
9. H. I. Bolker and N. Terashima, Infrared spectroscopy of
    lignins IV, "Lignin Structure and Reactions," Advan.
    Chem. Ser. 59, 110 (1966).
10. K. L. Paciorek, R. H. Kratzer, J. Kaufman, J. Nakaharg,
    and A. M. Hartstein, J. Fire Flammability 5, 243
    (1974).
11. F. Shafizadeh and Y. L. Fu, Carbohyd. Res 29, 113,
    (1973).
12. A. E. Lipska and F. A. Wodley, J. Appl. Polym. Sci. 13,
    851, (1969).
13. A. Broido and M. A. Nelson, Combustion Flame 24(2), 263,
    (1975).

INFLUENCE OF CRYSTALLINITY

ON THE THERMAL PROPERTIES OF CELLULOSE

K. E. Cabradilla
Department of Clothing, Textiles, and Interior Design
Kansas State University
Manhattan, Kansas

S. H. Zeronian

Division of Textiles and Clothing
University of California
Davis, California

I. INTRODUCTION

The pyrolysis of cellulose has been extensively studied
and much has been learned [1, 2]. However, little research
has been done on the effect of fine structure on the pyrolysis of cellulosic fibers. The most studied fine-structure
parameter is crystallinity. Previous studies on the influence of crystallinity of cellulose on its pyrolysis have
produced conflicting results. Results obtained, especially
from thermal analysis, are difficult to correlate due to
such differences existing between different types of amorphous cellulose as degree of polarization, carboxyl content,
carbonyl content, and proposed fine structure.

Examination of amorphous cellulose has been used to
study the influence of crystallinity on pyrolysis. Broido

and co-workers [3, 4] have reported that amorphous cellulose prepared by liquid ammonia treatment at -60°C for 4 hr and subsequent evaporation of the ammonia recrystallized upon subsequent heating. X-ray analysis showed that the amorphous cellulose regained its crystallinity when isothermally heated at 250°C. Subsequent heating caused a drop in crystallinity; x-ray analysis indicated a different cell form. The results of Patai and Halpern [5] showed that ball-milled filter paper, which possessed no crystallinity, did not "recrystallize" on heating. Other workers [6, 7] have reported that amorphous cellulose prepared by deacetylation of cellulose triacetate did not become crystalline during heating up to 190°C. Only the regularity of the molecular structure seemed to increase.

Thermal analysis of amorphous cellulose also shows conflicting results. Several workers [8-10] have reported DTA curves for amorphous cellulose similar to that of cellulose. However, the temperature at which the major pyrolytic decomposition occurs is lowered. These results indicate that amorphous cellulose decomposes more readily at lower temperatures than crystalline cellulose. On the other hand, an exothermic peak at 150°C for deacetylated triacetate has been reported [6, 7]. If this amorphous cellulose is oriented by drawing, the exothermic peak shifted to a higher temperature and became smaller. Also when amorphous cellulose was heated to 190°C, the peak disappeared completely. These workers suggested that this peak is specific to amorphous cellulose prepared from cellulose acetate by saponification. The char yield of amorphous cellulose has been reported to be greater than with crystalline cellulose [4, 11]. The higher char yield is thought to be dependent on the amount of amorphous material initially present in the cellulose.

The purpose of this research was to investigate how variation in the crystallinity of cotton cellulose affects its pyrolysis by studying a series of cotton celluloses of different crystallinities and fully amorphous cellulose prepared from commercial cellulose acetate.

II. EXPERIMENTAL

A series of cotton celluloses, having different amorphous fractions, was prepared. The crystallinities of these samples were determined by moisture regain measurements and x-ray diffraction. Moisture regain measurements were used to calculate the amount of amorphous material F_{am}, while x-ray diffraction was used to measure the crystallinity index CrI. The values for F_{am} and CrI for these cotton celluloses are given in Table I. These values do not give the same proportions of crystalline and amorphous cellulose, since x-ray diffraction and moisture regain measure different parameters of crystallinity [12]. Since methods are not yet available for making cotton cellulose fully amorphous without destruction of its fibrous character, deacetylated commercial diacetate filament yarn was used as a fully amorphous cellulosic material. Cotton celluloses, having different degrees of crystallinity, were characterized by degree of polymerization, copper number, and carboxyl content (Table I).

A. Materials

Partially purified cotton fiber, Deltapine-Smoothleaft variety of American upland cotton, obtained from the Southern Regional Research Center, USDA, and bright diacetate filament yarn (150-2Z-40) obtained from the Celanese Corporation was used. Cupriethylenediamine hydroxide solution was obtained from Ecusta Paper Division, Olin-Matheson Chemical Corp., Pisgah Forest, North Carolina. Other chemicals, except sodium methoxide, were reagent grade.

B. Methods of Treatment

The cotton fiber was further purified by refluxing for 8 hr in 1% sodium hydroxide (40 gm per 2 liter solution) in a nitrogen atmosphere, washed with distilled water until free from alkali, and then dried. All subsequent treatments were done on the purified cotton.

TABLE I

Characterization of Cotton Celluloses Having Different Degrees
of Crystallinity

F_{am} [a]	CrI	DP	Copper number	Carboxyl content (mM/100 gm)	Decrystalli- zation method
0.38	90.2	2070	0.01	0.43	None
0.52	75.9	1980	0.09	0.28	Ethylamine
0.59	70.6	847	0.79	0.20	Acetylation,then deacetylation with 0.1 N NaOCH$_3$
0.69	0	627	0.74	0.47	Acetylation,then deacetylation with alcoholic 1% NaOH
1.00 [b]	--	236	3.21	5.15	Deacetylation with alcoholic 1% NaOH

a. F_{am} = sorption ratio/2.6 [20].
b. Commercial diacetate filament yarn used as starting
material.

1. Acetylation

Cotton (10 gm) was soaked for 1 hr in distilled water,
filtered, and then soaked for 1 hr in glacial acetic acid
[13]. The cotton was filtered to remove excess acetic acid
before it was placed in a stoppered ground glass bottle
containing the acetylating solution. This consisted of 250
ml toluene, 125 ml acetic anhydride, and 0.05 ml 60% per-
chloric acid. Acetylation was allowed to proceed for 2 hr at
25°C. The sample was then washed periodically with distilled
water over an 18 hr period until it was acid free. Finally,
it was air dried.

2. Deacetylation with alcoholic 1% sodium hydroxide

Deacetylation of acetylated cotton and diacetate filament yarn was done according to the method of Jeffries [14].
The acetylated sample was soaked in alcoholic 1% sodium
hydroxide (1 gm per 100 ml solution) for 24 hr at room temperature. The sample was then washed with ethanol and kept
in ethanol for 18 hr. Finally, the product was air dried.

3. Deacetylation with 0.1 N sodium methoxide

Deacetylation of acetylated cotton was done according
to the method of Timell [15]. Acetylated cotton was shaken
with 0.1 N methanolic sodium methoxide (1 gm per 30 ml solution) at room temperature for 24 hr. Next, the sample was
washed with methanol, followed by methanol containing 5%
acetic acid and then again with methanol. Finally, the
product was air dried.

4. Decrystallization by ethylamine

Partial decrystallization of cotton by ethylamine was
obtained by the method described by Nevell and Zeronian [16].
Cotton was placed in a resin reaction kettle and the system
was evacuated. Ethylamine (10 gm per 400 ml solution) was
slowly added to the closed system. The reaction was allowed
to proceed for 4 hr at 0°C. To terminate the reaction, the
vacuum was released and the cotton was placed in distilled
water at 0°C. The sample was then washed in ice-cold distilled water over an 18 hr period. Finally, the product was
air dried.

5. Heat treatment of cotton of lowered crystallinity

Cotton deacetylated with alcoholic sodium hydroxide
(10 mg) was heated in the furnace of a Perkin-Elmer TGS-1
thermobalance. The samples had been ground in a Wiley mill
to pass a 20 mesh screen. After purging the system with
nitrogen, the sample was heated to the desired temperature
at a rate of 10°C/min. The sample was immediately cooled.

6. Levoglucosan (1,6-anhydro-β-D-glucopyranose

Levoglucosan was synthesized by the method of Coleman [17]. Penta-O-acetyl-β-D-glucopyranose was prepared by the acetylation of α-D-glucose [18]. Distillation of the pentacetate with phenol and p-toluenesulfonic acid produced phenyl tetra-O-acetyl-β-D-glucopyranoside [17]. Treatment of this product with boiling sodium hydroxide, followed by acetylation, yielded levoglucosan triacetate. Deacetylation with methanolic sodium methoxide yielded levoglucosan [19]. The levoglucosan was recrystallized from methanol. The melting point of the crystalline levoglucosan was 179°-180°C.

C. Characterization of Product

1. Hygroscopicity

Moisture regains were determined by drying duplicate samples (1.0 gm) of the materials over phosphorus pentoxide in an evacuated desiccator to a constant weight and then exposing at 59% R.H. (obtained by means of a saturated solution of sodium bromide) at 21°C until they reached a constant weight. In order to convert the moisture regains to sorption ratios, the moisture regain of the starting cotton was determined under the same set of conditions. Sorption ratio is defined as the ratio of moisture regain of a sample to that of the starting cotton at the same relative humidity and temperature. Amorphous fraction F_{am} was calculated according to Valentine's reation [20]:

$$F_{am} = \text{sorption ratio}/2.6.$$

2. X-ray analysis

Diffractograms were taken of fibers, which had been cut into 1/8 in pieces, with a high-angle precision x-ray diffractometer, using nickel-filtered CuKα= radioation. Crystallinity indices CrI were calculated from the diffractograms using the empirical relation of Segal et al [21].

3. Degree of polymerization

The intrinsic viscosity of the samples were determined with Cannon-Fenske viscosimeters in cupriethylenediamine hydroxide, using ASTM D1795-62 [22]. Intrinsic viscosity was converted to weight average degree of polymerization (DP) using the relation:

$$DP = 190 \ [7]$$

4. Copper number

The copper number of the samples were determined using Braidy's method [23].

5. Carboxyl content

Carboxyl content of the samples were determined using the methylene blue method of Davidson [24].

6. Differential thermal analysis

Differential thermal analysis (DTA) was run on samples (4 mg) in a nitrogen atmosphere on a Deltatherm III differential thermal analyzer. Fibrous samples were cut into 1/4 inch pieces to ensure uniform packing and good reproducibility and then packed into aluminum liner pans. Samples that had been previously ground in a Wiley mill to pass a 20 mesh screen were packed into aluminum liner pans in their original condition. The system was purged with nitrogen before heating. The heating rate was 10°C/min. Alumina was used as the reference material. Each sample was run four times. To analyze the DTA curve, two measurements of temperature were calculated, T_{onset} and T_{peak} (Fig. 1). T_{onset} is the temperature at the intersection of the extrapolated base line and the extrapolated line of the endotherm or exotherm. T_{peak} is the temperature at the insection of the extropolated lines of the endotherm or exotherm.

7. Thermogravimetry

Thermogravimetry (TG) was run on the samples (4 mg) in a nitrogen atmosphere on a Perkin-Elmer TGS-1 thermobalance. Fibrous samples were cut into 1/4 inch pieces and then

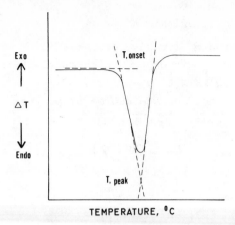

TEMPERATURE, °C

Fig. 1.   Analysis of differential thermal
analysis (DTA) data.

packed in aluminum pans.   The system was purged with nitrogen
before heating.   The heating rate was either 5°C/min or
16°C/min.   Each sample was run three or four times.   Temperature at which the weight loss reaction started (T), percent
residue for the reaction, and percent char were calculated
from the TG curve (Fig. 2).   Temperatures were corrected by
subtraction of a correction factor (obtained from a calibration curve) from the recorded temperature.

Derivative thermogravimetry (DTG) was run on samples
(4 mg) under nitrogen on a Perkin-Elmer TGS-1 thermobalance.
Fibrous samples were cut into 1/4 inch pieces and then
packed in aluminum pans.   The system was purged with nitrogen before heating.   The heating rate was 160°C/min.   Each
sample was run twice.   The temperature where the maximum
rate of change of weight occurred was obtained from the
curves.

8.   Levoglucosan analysis

Gas chromatograph (GC) was used to analyze the amount
of levoglucosan in the silylated tar of pyrolyzed cotton
samples.   Fibrous samples (0.5 gm) were cut into 1/4 inch
pieces for pyrolysis.   Samples were dried for 3 hr at 110°C
and then weighed to determine their dry weight.   The samples

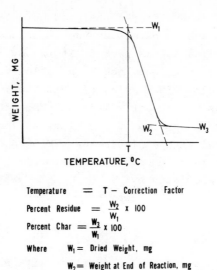

$$\text{Temperature} \quad = \quad T - \text{Correction Factor}$$

$$\text{Percent Residue} \quad = \quad \frac{W_2}{W_1} \times 100$$

$$\text{Percent Char} = \frac{W_3}{W_1} \times 100$$

$$\text{Where} \quad W_1 = \text{Dried Weight, mg}$$

$$W_2 = \text{Weight at End of Reaction, mg}$$

$$W_3 = \text{Weight at 500 °C, mg}$$

Fig. 2.   Analysis of thermogravimetric TG curve.

were packed in combustion boats and placed in the pyrolysis apparatus.

The pyrolysis apparatus consisted of a glass tube (1 inch diameter, 24 inches long) that was fitted in a tube furnace (8 inches long). A sample was placed in the apparatus and the system was evacuated for 10 min. Then the portion of the tube that contained the sample was moved into the center of the furnace. The sample was pyrolyzed for 1 1/2 hr under a vacuum of 1.5 mm Hg at 350°C. The tar collected on the cool part of the tube. To terminate the pyrolysis, the vacuum was released and the tube removed from the furnace. The pyrolyzed sample was weighed to determine the percent residue. The percent residue is the weight of residue left after pyrolyzing the samples for 1 1/2 hr at 350°C divided by the dry weight of the sample. The tar was removed from the tube by washing it with 10 ml of pyridine containing p-terphenyl (0.0625 gm/10 ml) as the internal standard. The tar was silylated by adding 0.5 ml of the tar solution to 1 ml of Tri-Sil Z and heating at 60°-70°C for 3 min.

The silylated tar was analyzed on a Varian Model 1200
gas chromatograph to determine the levoglucosan content.
A 3 m x 2 mm i.d. column packed with 2% OV-17 on Chromosorb G,
AW-DMCS, 70/80 mesh, was used. Variables were: carrier gas
and hydrogen flow, 30 ml/min, and air flow, 150-300 ml/min.
The column was heated after injection at 2°C/min from 150°C
to 270°C. The peak area of levoglucosan and p-terphenyl
were determined with a planimeter. From a calibration curve,
the weight of levoglucosan was determined. The levoglucosan
content was calculated by:

$$\text{Levoglucosan content} = \frac{\text{weight of levoglucosan in tar}}{\text{dry weight of cellulose.}} \times 100$$

3. Infrared analysis

Infrared spectra were obtained with a Perkin-Elmer
IR-257 grating spectrophotometer by use of the potassium
bromide pellet technique. Measurement of crystallinity was
calculated from the infrared ratios, $a_{1372\ cm^{-1}}/a_{2900\ cm^{-1}}$
and $a_{1429cm^{-1}}/a_{893\ cm^{-1}}$ [25, 26].

III. RESULTS AND DISCUSSION

A. Thermal Analysis

Thermal analysis of cotton celluloses, having different
degrees of crystallinities, showed tremendous differences as
the F_{am} increased. The DTA curve of cotton cellulose
(F_{am} of 0.38) showed a major endothermic decomposition beginning at 332°C and peaking at 358°C (Fig. 3). The thermal
decomposition endotherm moved to lower temperatures as the
F_{am} increased from 0.38 to 0.59. This decrease in the temperature can be attributed simply to the lower DP for cotton
cellulose with F_{am} of 0.52 [27]. For the sample with F_{am} of
0.59, the lowered decomposition temperature appears to be
caused by the decrease in DP, increasing in reducing power,
and the increase in the amorphous fraction. A similar
decrease in the temperature of the major pyrolytic decomposition for amorphous cellulose was found by other workers
[8-10]. However, when the F_{am} was increased to 0.69, there

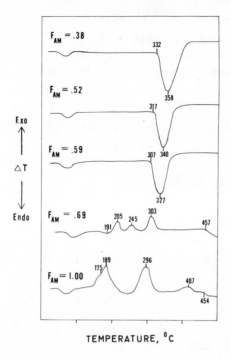

TEMPERATURE, ^{0}C

Fig. 3. Differential thermal analysis
(DTA) curves of cotton celluloses having
different degrees of crystallinities heated
at 10°C/min.

was no longer a major endothermic decomposition peak.
Instead there was a series of exothermic peaks between 180°-
320°C: an exotherm starting at 191°C that peaks at 205°C,
followed by a milder exotherm peaking at 245°C, and a third
exotherm peaking at 303°C. Additionally, there was a broad
endotherm beginning at 457°C. This drastic change in the DTA
curve of cotton cellulose with F_{am} of 0.69 indicates that
there is a change in the type of pyrolysis reaction occurring.
For deacetylated acetate (F_{am} of 1.00), a similar series of
exotherms was also observed. There was a large exotherm
peaking at 189°C that had an exothermic shoulder at 175°C, a
second exotherm peaking at 296°C, a slight endotherm at 407°C,
and a broad endotherm starting at 454°C.

TG also showed the effect of crystallinity on the pyrolysis of cotton cellulose. A typical TG curve for cotten (F_{am} of 0.38) showed a slight loss due to elimination of physically absorbed water below 100°C, a very slight loss in weight between 100°-280°C, and major weight loss due to thermal decomposition between 280°-380°C (Fig. 4). There was a slight weight loss between 380° and 500°C, indicating that there was further reaction involving the char. When the F_{am} of cotton was increased from 0.38 to 0.59, the TG curve retained the same general features. But there was a slight

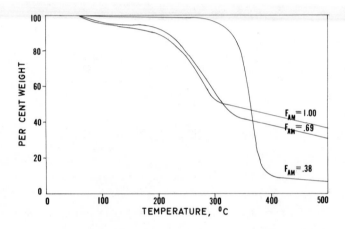

Fig. 4. Thermogravimetric (TG) curves of cotton celluloses having different degrees of crystallinities heated at 5°C/min.

lowering in the temperature at which weight loss start (Table II). The percent char and percent residue remained unchanged with this increase in F_{am} from 0.38 to 0.59. For cotton cellulose with F_{am} of 0.69 and for deacetylated cellulose acetate (F_{am} of 1.00), the temperature at which the major weight loss begins fell by 100°C (Table II, Fig. 4). In the case of these samples, a gradual weight loss started after 100°C. The weight rapidly began to decrease at approximately 200°C and continued to approximately 330°C, where there was a gradual weight loss to 500°C. The major weight loss portion of the TG curves for cellulose samples having

TABLE II

Thermogravimetric Data of Cotton Celluloses Having Different
Degrees of Crystallinity

F_{am} [a]	Temperature ($^\circ$C) [c]	Residue (%)	Char (%)
0.38	337	11.7	8.9
0.52	325	12.4	9.5
0.59	314	11.3	7.2
0.69	223	44.6	33.5
1.00 [b]	215	49.6	36.6

a. F_{am} = sorption ratio/2.6 [20].

b. Commercial diacetate filament yarn used as starting
material.

c. Onset of weight loss; heating rate = 5°C/min.

F_{am} of 0.69 and 1.00 was not linear, as in the case of cotton
(F_{am} of 0.38). This portion of the TG curve appeared to be
slightly concave, indicating that two reactions may be occurring, one immediately following another. The percent of
residue and percent char increased by a factor of 4 for these
samples of high F_{am}, compared to samples of lower F_{am}
(Table II).

DTG showed that the rate of weight change is affected
by crystallinity. A faster heating rate (160°C/min) was used
to measure the rate of weight change, since preliminary
results at a heating rate of 5°C/min did not show the rate of
weight change. This was due to the fact that the rate of
weight change was measured in milligrams per second. Since
earlier TG measurements (Table II, Fig. 4) had been done at
a heating rate of 5°C/min. additional TG determinations were
made at a heating rate of 160°C/min to quantize the effect
of the heating rate. When cotton cellulose was heated at a
fast rate (160°C/min) (Table III), the temperature at which
the weight loss occurred shifted to a higher temperature
when compared to a slower heating rate (5°C/min) (Table II).
This higher decomposition temperature was due to the faster

TABLE III

Thermogravimetric and Differential Thermogravimetric Data of
Cotton Celluloses Having Different Degrees of Crystallinity
Heated at 160°C/min.

F_{am} [a]	Thermogravimetry temperature (°C) [b]	Differential Thermogravimetry temperature (°C) [c]
0.38	403	456
0.52	390	448
0.59	375, 422	412, 437
0.69	269, 326	305, 360
1.00 [d]	270, 323	323, 336

    a. F_{am} = sorption ratio/2.6 [20].

    b. Temperature at which weight loss begins (Fig. 2).
Where two readings are given, the second indicates the position of an inflection point.

    c. Temperature where the rate of change of weight
reaches its maximum value. Where two readings are given, the
first indicates the position of a shoulder on the curve.

    d. Commercial diacetate filament yarn used as a start-
ing material.

heating rate; the sample temperature cannot catch up to the
environmental temperature, and therefore, the recorded
temperature was higher than the sample temperature [2]. When
the F_{am} was increased to 0.59, the weight loss portion of
the TG curve (heating rate of 160°C/min) was no longer
linear, but was slightly concave. Two temperatures were
measured from the TG curves; the second temperature measured
was the temperature where the slope of the weight loss portion of the TG curve changes (Table III). This change in the
shape of the TG curve was due to the faster heating rate and
temperature lag. As F_{am} was increased to 1.00, the rapid
weight loss portion of the TG curve became more concave. For

samples of F_{am} of 0.69 and 1.00, there was a gradual weight
loss beginning at 120°C, which was not present for samples
with lower F_{am}; this weight loss was greater with the samples
having F_{am} of 1.00.

DTG curves showed that the temperature where an inflection occurs in the rate of weight loss curve (i.e., the
derivative thermogravimetric curve) was lowered with a
decrease in crystallinity (Table III, Fig. 5). As the F_{am}
was increased from 0.38 to 1.00, the temperature where the
maximum value of the rate of change of weight occurred
decreased from 456°C to 336°C. When F_{am} was increased to
0.59, a shoulder appeared on the rate peak, indicating two
inflections in the rate of weight loss curve. This shoulder
became more prominent as the F_{am} was increased to 0.69 and
still more prominent at a F_{am} of 1.00. For samples having

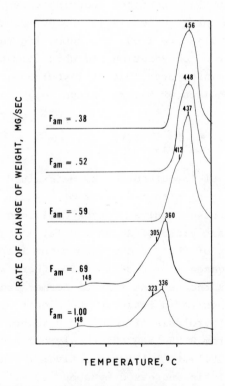

Fig. 5. Derivative thermogravimetric (DTG) curves of
cotton celluloses having different degrees of crystallinities.

F_{am} of 0.69 and 1.00, the DTG curve showed a slight weight loss from 148°C to where the major rate of weight change began at approximately 275°C.

The pyrolysis of cellulose involves two competing primary decomposition pathways [1, 2]: (a) levoglucosan formation, which occurs above 300°C; (b) dehydration, which occurs below 300°C, followed by char formation. The results of thermal analysis of cotton cellulose used in this study were similar to those of other investigators [28-30]. The major endothermic decomposition occurring between 280°C and 380°C, which leaves little char, is due to the formation of levoglucosan and its subsequent reactions. When the amorphous fraction is increased to 0.59, levoglucosan formation still remained the major decomposition pathway. However, thermal decomposition occurred at lower temperatures due to both decreased DP and increased reducing power and due to increased amorphous fraction [27]. However, when F_{am} was increased to 0.69 and to 1.00, the competing thermal decomposition pathway of dehydration and char formation appeared to become predominate over that of levoglucosan formation. This pathway consists of several steps, each involving the loss of weight.

B. Levoglucosan Analysis

The yield of levoglucosan was determined for cotton celluloses having different degrees of crystallinities. Table IV lists these results. Cotton cellulose (F_{am} of 0.38) had a levoglucosan yield of 28.5% and 13.5% char residue after pyrolysis of 350°C under vacuum for 1 1/2 hours. As the F_{am} increased from 0.38 to 0.59, there was essentially no difference in the percent residue formed from the pyrolysis and a progressive increase in the amount of levoglucosan produced in the pyrolysis of cotton. However, when the F_{am} was increased from 0.59 to 0.69, the levoglucosan yield decreased from 39.1% to 1.1%, respectively, and the residue left from pyrolysis increased by about 300%. This drastic lowering in the amount of levoglucosan produced and increased production of char indicate a change in the course of

TABLE IV

Levoglucosan Content of Pyrolyzed Cotton Celluloses Having
Different Degrees of Crystallinity

F_{am}[a]	Levoglucosan[b] (%)	Residue[b] (%)
0.38	28.5	13.5
0.52	31.0	14.0
0.59	39.1	12.5
0.69	1.1	41.7
1.00[c]	0.2	41.8

a. F_{am} = sorption ratio/2.6 [20].
b. Based on dry weight of cellulose. Residue is the
residue left after pyrolysis to determine the levoglucosan
content.
c. Commercial diacetate filament yarn used as starting
material.

pyrolysis. The yield of levoglucosan and the percent char
from cotton with F_{am} of 0.69 and deacetylated cellulose
acetate with F_{am} of 1.00 were essentially the same. It is
improbable that changes in chemical composition of the cellulose as reflected by copper numbers, which were similar for
cotton with F_{am} of 0.59 and with F_{am} of 0.69; DP, which
differed by 200; and carboxyl content could account for the
drastic drop in levoglucosan formation as F_{am} increased from
0.59 to 0.69. Also, the chemical composition of the fully
deacetylated acetate (F_{am} of 1.00) does not appear to affect
the amount of levoglucosan formed since the levoglucosan
content of pyrolysis products for this sample was similar to
that of cotton with F_{am} of 0.69.
    Crystallinity appeared to have a significant effect on
the pyrolysis of cotton cellulose. The extent to which the
reaction pathway of levoglucosan formation competes with that
of dehydration and char formation may depend on the degree of

crystallinity, as indicated by either x-ray data (CrI) or sorption ratio ($1 - F_{am}$).

For cotton cellulose ($F_{am} = 0.38$, CrI = 90.2), the major pyrolytic reaction is that of levoglucosan formation; this was shown by the major endothermic decomposition (Fig. 3) and levoglucosan content (Table IV). As the degree of crystallinity was decreased ($F_{am} = 0.59$, CrI = 70.6), the formation of levoglucosan was still the major pyrolytic reaction. Since two peaks occurred in the DTG curve (Fig. 5), it appeared that an additional reaction had also occurred. However, levoglucosan content and thermal analysis indicated that the reaction of levoglucosan formation was the major reaction at this level of crystallinity. When the degree of crystallinity was lowered further ($F_{am} = 0.69$, CrI = 0), the formation of levoglucosan was no longer favored. It was of minor importance, as shown by the low levoglucosan content (Table IV). The major pyrolytic reactions were those of dehydration and char formation. This change in the course of pyrolysis was shown by the alteration in the DTA curve, in which there was no longer an endothermic decomposition (Fig. 3). Dehydration and char formation were still the predominate reactions for the completely amorphous cellulose prepared from deacetylated acetate. Thus, two types of reactions can occur during the pyrolysis of cotton cellulose: levoglucosan formation and dehydration and char formation. These reactions do not occur in regions of similar supramolecular structure. Levoglucosan formation was favored in cotton cellulose ($F_{am} = 0.38$, CrI - 0.92) and in samples with F_{am} of 0.6 and lower while dehydration and char formation were favored in samples with F_{am} of 0.7 and higher. There is a larger amount of crystalline material than amorphous material in cotton. As the degree of crystallinity was reduced to low levels, dehydration and char formation predominated over levoglucosan formation. Thus, levoglucosan formation was favored in the crystalline regions, while dehydration and char formation were favored in the amorphous or disordered regions; the extent to which these competing types of reactions occurred was controlled by the degree of crystallinity.

The major pyrolytic reaction in the amorphous regions of cellulose was that of dehydration. This reaction involved the loss of a molecule of water. After dehydration, there was subsequent breakdown of the "anhydrocellulose" or "dehydrocellulose" [31-36]. This was indicated by thermal analysis (Fig. 3-5). For cotton cellulose with F_{am} of 0.69, the first step was dehydration and was shown by the exotherm, which started at 191°C and peaked at 205°C. This step was completed before the start of the second step of pyrolysis, which was represented by the exotherm peaking at 245°C. This second step has been suggested to involve the further degradation of the "anhydrocellulose" or "dehydrocellulose" [36]. The final step again was exothermic, peaking at 305°C. This may be due to depolymerization and levoglucosan formation [27]. The sequence of steps involving dehydration was finished before levoglucosan formation (a minor reaction in this case) began. TG and DTG also confirmed this multistep pathway. Amorphous cellulose (F_{am} = 2.00), prepared by deacetylation of commercial cellulose diacetate, exhibited a similar dehydration reaction. However, the steps involved in dehydration overlapped, as indicated by DTA (Fig. 3).

C. Effect of Preheating Cellulose

Some investigators [3, 4] have suggested that crystallization occurs when decrystallized or amorphous cellulose is heated. It is possible that crystallization may be occurring simultaneously with thermal decomposition. Cotton cellulose with F_{am} of 0.69 was heated in order to determine if changes in crystallinity occurred during heating.

Cotton cellulose with F_{am} of 0.69 showed differences in its infrared spectrum when heated to various temperatures and immediately cooled (Fig. 6). When cotton cellulose (F_{am} = 0.69) was heated to 150°C and immediately cooled, the absorption band at 1655 cm^{-1} was stronger. Also, a new absorption band at 810 cm^{-1} appeared; this may be due to olefinic C-H bending vibrations. Upon heating to 200°C, the absorption band at 1655-1660 cm^{-1} weakened and a new band appeared at 1600-1590 cm^{-1}. This new band at 1600-1590 cm^{-1}

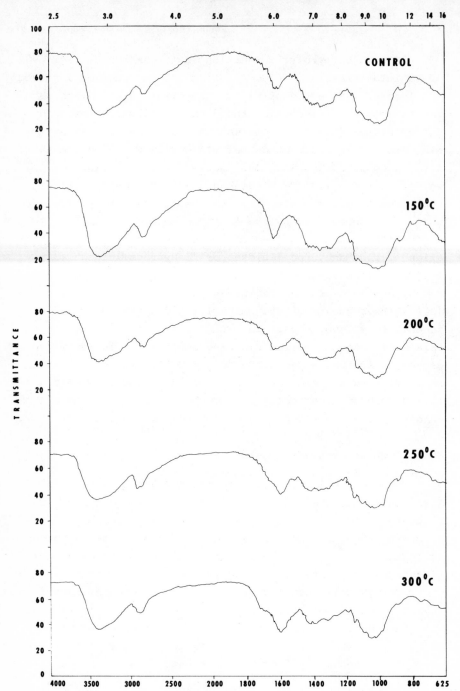

Fig. 6.   Infrared spectra of cotton cellulose with F_{am} of 0.69 after heating at various temperatures.

was previously reported by other workers [6, 7] in their
work on the effect of heat on deacetylated triacetate; they
attributed this band to hydroxyl groups, which tend to form
hydrogen bonds in amorphous cellulose. Also the band at
810 cm^{-1} had decreased in intensity. When the sample was
heated to 250°C and immediately cooled, the bands at
1660 cm^{-1} and 1625 cm^{-1} had disappeared and a new band
appeared at approximately 1700 cm^{-1}. There was no longer an
absorption band at 810 cm^{-1}. At 300°C, the absorption band
at 1700 cm^{-1} became more pronounced. The band at 839 cm^{-1}
had decreased in intensity.

There was no increase in the crystallinity in cotton
cellulose with F_{am} of 0.69 when heated to a given temperature
and immediately cooled. The change in crystallinity can be
measured by two infrared crystallinity or lateral order
indices, $a_{1372\ cm^{-1}}/a_{2900\ cm^{-1}}$ and $a_{1429\ cm^{-1}}/a_{893\ cm^{-1}}$
[25, 26]. The absorption bands at 1372 cm^{-1} and 1429 cm^{-1}
did not increase in intensity with heat treatment, but
remained similar to the control sample. Infrared crystallinity indices were not calculated for this sample, since the
absorption at 1372 cm^{-1} and at 1429 cm^{-1} was very small.

Thermal analysis of cotton cellulose with F_{am} of 0.69
showed differences as the heat treatment temperature was
increased. DTA curves were similar to the control when
heated to 150°C and to 200°C (Fig. 7). At 250°C, the first
exotherm of 205°C had disappeared. At 300°C, only the
single exotherm at 297°C and the broad endotherm beginning
at 448°C remained.

These changes in the DTA curves when cotton cellulose
with F_{am} of 0.69 was heated do not reflect a change in
crystallinity. If crystallization had occurred for this
cellulose sample, thermal analysis would have indicated this
by the appearance of an endothermic decomposition peak
between 300°C and 370°C, provided that crystallization
occurred before the formation of "anhydrocellulose" or
"dehydrocellulose." However, thermal analysis showed that
dehydration, not crystallization, occurred, since there was
no endothermic peak for samples heated between 150°C and
300°C. Changes in the DTA curves of cotton cellulose with

F_{am} of 0.69 heated up to 300°C also indicate the stepwise sequence of events in the pyrolysis reaction involving dehydration. It also appears likely that the steps in the dehydration reaction are not reversible.

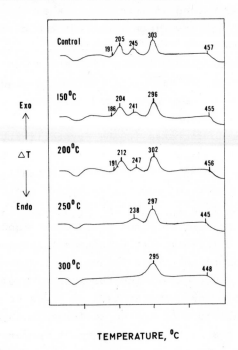

Fig. 7. Differential thermal analysis (DTA) curves of cotton cellulose with F_{am} of 0.69 heating to different temperatures.

ACKNOWLEDGMENT

The authors thank Mr. J. J. Creely, Southern Regional Research Center, USDA, New Orleans, Louisiana, for taking the x-ray diffractograms of the cellulose samples.

REFERENCES

1. F. Shafizadeh, Advan. Carbohyd. Chem. 23, 419 (1968).

2. P. K. Chatterjee and R. F. Schwenker, Instrumental
    methods in the study of oxidation, degradation, and
    pyrolysis of cellulose, in "Instrumental Analysis of
    Cotton Cellulose and Modified Cotton Cellulose"
    (R. T. O.'Connor, ed.) New York 1972.

3. M. Weinstein and A. Broido, Combustion Sci. Technol. 1,
    287 (1970).

4. A. Broido, A. C. Javier-Son, A. C. Ouano, and E. M.
    Barrall, J. Appl. Polym. Sci. 17, 3627 (1973).

5. S. Patai and Y. Halpern, Israel J. Chem. 8, 655 (1970).

6. H. Hatekeyama and T. Hatekeyama, Sen-i Gakkaishi 30,
    T214 (1974).

7. H. Hatekeyama,   T. Hatekeyama, and J. Makano, Cellu-
    lose Chem. Technol. 8, 495 (1974).

8. K. Kato and K. Komorita, Agric. Biol. Chem. (Tokyo) 32,
    21 (1968).

9. M. Kosik, V. Reiser, and I. Michlik, Vyskum Prace
    Odboru Papier Cellulozy 5, 29 (1973).

10. T. Yurugi, H. Iwata, and M. Okuma, Nippon Kagaku Kaishi
    8, 733 (1975).

11. A. E. Lipska and G. E. McCasland, J. Appl. Polym. Sci.
    15, 419 (1971).

12. J. O. Warwicker, R. Jeffries, R. L. Colbran, and R. N.
    Robinson, Shirley Inst. Pamph. No. 93, 1966.

13. K. E. Cabradilla and S. H. Zeronian, J. Appl. Polym. Sci.
    19, 503 (1975).

14. R. Jeffries, J. Appl. Polym. Sci. 12, 425 (1968).

15. T. E. Timell, J. Polym. Sci., Part C, 2, 109 (1963).

16. T. P. Nevell, and S. H. Zeronian, Polymer 3, 187 (1962).

17. G. H. Coleman, 1,6-Anhydro-β-D-glucopyranose triacetate
    (levoglucosan triacetate), in "Methods in Carbohydrate
    Chemistry" Vol. II (R. L. Whisler, ed.). Academic
    Press, New York, 1963.

18. M. L. Wolfrom and A. Thompson, Acetylation in "Methods
    in Carbohydrate Chemistry," Vol. II, (R. L. Whisler,
    ed.). Academic Press, New York, 1963.

19. R. B. Ward, 1,6-Anhydro-β-D-glucopyranose (levoglucosan)
    in "Methods in Carbohydrate Chemistry," Vol. II
    (R. L. Whisler, ed.). Academic Press, New York, 1963.

20. L. Valentine, Chem. Ind. (London), 1279 (1956).

21. L. Segal, J. J. Creely, A. E. Martin, and C. M. Conrad,
    Textile Res. J. 29, 786 (1959).

22. "ASTM Standards, Part 15." American Soceity for Testing
    Materials, Philadelphia, 1970.

23. T. P. Nevell, Determination of reducing end-groups, in
    "Methods in Carbohydrate Chemistry" Vol. III (R. L.
    Whistler, ed.). Academic Press, New York, 1963.

24. O. Samuelson, "Determination of carboxyl groups, in
    "Methods in Carbohydrate Chemistry," Vol. III
    (R. L. Whistler, ed.). Academic Press, New York, 1963.

25. R. T. O'Connor, E. F. Du Pre, and D. Mitcham, Textile
    Res. J. 28, 382 (1958).

26. M. L. Nelson and R. T. O'Connor, J. Appl. Polym. Sci. 8,
    1325 (1964).

27. K. E. Cabradilla, Ph. D. Thesis, University of Califor-
    nia, Davis, 1976.

28. K. Akita and M. Kase, J. Appl. Polym. Sci., Part A-1 5,
    833 (1967).

29. F. Shafizadeah and Y. L. Fu, Carbohyd. Res. 29, 113
    (1973).

30. R. M. Perkins, G. L. Drake, and W. A. Reeves, J. Appl.
    Polym. Sci. 10, 1041 (1966).

31. S. L. Madorsky, Cellulosic materials, in "Thermal
    Degradation of Organic Polymers," (S. L. Madorsky, ed.)
    Wiley (Interscience), New York, 1964.

32. D. F. Arseneau, Can. J. Chem. 49, 632 (1971).

33. K. S. Patel, K. C. Patel, and R. D. Patel, Makro. Chem.
    132, 7 (1970).

34. F. J. Kilzer and A. Broido, Pyrodynamics 2, 151 (1965).

35. W. Hoffman, T. Ostrowski, T. Urbanski, and M. Witanowski,
    Chem. Ind. (London) 95 (1960).

36. M. M. Tang and R. Bacon, Carbon 2, 211 (1964).

THE PYROLYSIS-GASIFICATION-COMBUSTION PROCESS

ENERGY CONSIDERATIONS AND OVERALL PROCESSING

David L. Brink, James A. Charley

George W. Faltico, and Jerome F. Thomas

Forest Products Laboratory
University of California
Berkeley, California

I. INTRODUCTION

In 1792, Robert Mudoch introduced fuels produced by pyrolyzing wood. Coal quickly superseded wood for the production of a synthetic gas. Then coal was surpassed by petroleum and natural gas--still more convenient energy sources. During the past five years the alarming rate of depletion of domestic petroleum and natural gas reserves became generally recognized. Increasing attention once again has been placed on systems for conversion of coal and, indeed, plant materials to clean, convenient energy fuels. Pyrolysis is again being considered as a technique for recovering energy from wood and other plant materials. Moreover, these renewable materials can provide significant supplies of environmentally desirable fuels.

An accessible and economically beneficial source of
plant material constitutes the bulk of this nation's solid
waste output. In the United States, over 129 million tons of
raw solid wastes are generated annually by urban sources [1].
Urban solid waste, received with an average 21% moisture and
25% ash content, has a net heating value of roughly 10 million BTUs per ton [2]. This represents some 1.25 quadrillion
BTUs (quads) per year. Additionally, it has been conservatively estimated that organic wastes from industrial, agricultural, and forestry sources contribute six times this
amount; approximately 750 million tons annually. With development of appropriate management techniques, it should be
possible to collect a substantial portion of this resource
and thereby increase the potential for power generation
several fold.

Such solid waste conversion would enhance the quality
of our environment through the reduction of water pollution
caused by ocean and river dumping and landfill leaching, land
alteration from landfills or open dumps, and air pollution
from uncontrolled burning of various wastes.

Greater use of the earth's biomass for generation of
energy has lately aroused attention. Aspects of this
resource have been considered previously [3] together with
the use of organic residues. By capturing and storing solar
energy through carbon fixation, biomass converted to combustible fuels has formed one of our most convenient and most
used sources of energy, at least since man began to smelt
metals in about 3500 B.C. Photosynthesis renews the biomass
with a yearly fixation of roughly 150 billion tons of carbon.
Calculated at 14,000 BTUs per pound, this source of carbon
provides thirty times more energy than the current total
annual world energy consumption of 144 quads. Thus, efficient conversion of a small fraction of the biomass could
provide a substantial energy source.

Combustion, pyrolysis-gasification, anaerobic fermenta-
tion, enzymatic hydrolysis, or more exotic techniques may
supply the United States with between 3 quadrillion BTUs [4]
and 15 quadrillion BTUs [5] annually by the year 2000.
Efforts to increase biomass production and use could reduce

dependence on seriously polluting, nonrenewable fuels. Such efforts would provide a salubrious effect on the earth's overall environment. Reclamation of unproductive and under-stocked forest land and, with adequate supplies of water, range lands and even desert lands would be enhanced. New economic growth and development would be viable in otherwise disadvantaged regions.

To maximize returns, biomass production and harvesting would demand application of enlightened management tech-niques. These would prevent undesirable modifications or damage to existing ecosystems. With comprehensive management, biomass cultivation can be directed towards maximizing production of food and material products in high demand such as sugars, proteins, fiber or fodder. Once the primary product (i.e., food) is extracted or removed, residues such as cornstalks or logging residues could be reclaimed for their energy content. Such planned harvesting raises the status of the energy product from a byproduct to a co-product. Ultimately, matter remains conserved; consumer products from the biomass harvested as materials will eventu-ally find their way to the organic waste stream.

Perhaps, then, the major obstacle, assuming adequate interest, is lack of bioconversion systems proven to be both economically rewarding and environmentally sound. Published evaluations of systems relative to capital and operating (especially maintenance) costs for full scale conversion have not been attractive. In several instances, environmental compatibility has proved to be a problem. Indeed, before this decade energy appeared to be too plentiful and cheap to place really significant efforts on development of such sys-tems. Now man continues to reach ever deeper into the earth to burn prehistoric reserves of stored solar energy--fossil fuels. He risks the imponderable problems of plutonium proliferation. A sound future demands alteration of energy supply and consumption patterns. Even if still not justified by economic accountability, increased support of research and development of energy recovery from renewable plant matter and conservation of our remaining fossil resources is a moral responsibility.

II. DIRECT COMBUSTION OF PLANT MATERIAL

Direct combustion of plant material to satisfy man's energy needs dates back to prehistoric time. Even so, some of the same problems in its use persist today: air pollution and poor overall efficiency. The forest products industry has been in the forefront of implementing improved methods to burn plant materials. Initial designs concerned simple means for producing low pressure steam or the cheapest way to dispose of wood and bark residues. As fuel concerns increased, efficient incinerators with waste heat boilers became standard equipment. Further developments borrowed from coal combustion technology to provide integrated fire-box-boilers with spreader stokers or cyclonic suspension firing. However, such efficient firing methods demanded corresponding advances in fuel preparations: pulverization, drying, and handling. In the last decade, pressing environmental concerns required increasingly sophisticated means for assuring complete combustion and removal of particulates from the flue gases.

There are still many disadvantages to be overcome. Grantham [7] points out boilers designed for firing wood operate at only 65% efficiency compared to 85% for fossil fuels. Furnace volumes must be greater, resulting in three to four times the capital cost for an installed boiler having the same efficiency. Additional equipment must be provided for solid fuel processing and handling with its attendant combustion problems. Air pollution controls are expensive. Combustion of plant material requires more excess air, and produces more particulate pollution than any other fuel. Finally, the direct combustion of green wood or refuse at a high moisture content is inefficient. Problems of air pollution are incurred when these kinds of wet materials are dried, or partially dried, resulting from the emission of volatile organic components.

Now municipalities are also showing renewed interest in building energy recovering solid waste incinerators as available landfill sites diminish. Unlike the burning of wood residue, however, combustion of municipal solid waste

results in excessive corrosion of boiler tubes at steam
temperatures over 300°C [6]. The decomposition of plastic
fractions produces chlorides while the decomposition of
proteinaceous and other material produces sulfur oxides.
These obstacles can be overcome by either generating low
pressure steam exclusively, supplementing coal streams in
existing steam electricity generating plants with up to 20%
solid waste, or adopting a two-stage boiler with a waste-
fired low pressure unit and a fossil-fuel fired high pressure
unit. However, such approaches do not maximize the production of power from the material available.

III. PYROLYSIS-GASIFICATION

Pyrolysis-gasification of organic materials has been
under development for the past fourteen years at the University of California Forest Products Laboratory. Certain
problems inherent in direct combustion have been circum-
vented by physically separating the stages of conventional
combustion and pyrolysis-gasification into two steps.

Pyrolysis can best be defined as the destructive distillation of organic material heated to 600°C without the
addition of oxygen. Rate of temperature change, residence
time, and chemical composition of material introduced into
the reactor determine the nature of the products formed. With
a slow temperature change, decompositon proceeds in an
orderly arrangement, forming increasingly stable molecules,
richer in carbon, converging toward graphite. Charcoal
becomes the primary product, with a heating value of between
11,000-14,000 BTU/lb. Very rapid heating, or "flash" pyrolysis, produces char and relatively long chain hydrocarbon
gases that, if not subject to further thermal degradation,
can be condensed into liquid fuels having heating values in
the 10,000-11,000 BTU range. Complete gasification, on the
other hand, occurs as the products of pyrolysis are heated
from 600°C to around 1000°C. Vapors and gaseous compounds
undergo further fragmentation and react with the remaining
carbonaceous residues. Virtually the only products then
remaining are carbon monoxide, water vapor, carbon dioxide,

hydrogen, methane, and an inorganic ash. The relative proportions of each gas are determined through altering the
ultimate reaction temperature, the mode of heating, the
gases used for heating, and the moisture content and composition of the feedstock. As such, the heating value of the
gas will range from 100 ro 400 BTUs per cubic foot.

Our work, which has involved laboratory bench-scale
processing of several hundred grams of feed stock per hour
and a large experimental unit-scale processing at nominal
rates up to one ton per day of solids or 2.5 tons per day of
kraft black liquor was recently reviewed with appropriate
references being given. In particular, results obtained on
the effect of feed stocks having widely varying compositions
were discussed. It was shown that moisture content was the
principal parameter affecting gas composition. At moisture
contents of 35 to 55%, kraft black liquor having a well
defined water soluble ash content of 39 to 47%, municipal
solid waste having an ill defined, largely water insoluble
ash of approximately 42%, and wood having an ash content of
0.3%, yielded very similar gaseous products as temperature
approached 1000°C. Cracking of complex organic vapors as
well as gasification of carbonaceous residues approached
completion. The resultant gaseous products were essentially
reduced to simple mixtures of H_2, CO, CO_2, H_2O, CH_4, and
when sulfur was present, hydrogen sulfide. Trace amounts of
ethane and ethylene were occasionally found. When water content of the feed was drastically reduced, the ultimate gas
products were qualitatively the same but molar compositions
were substantially different.

Principal chemical reactions, accounting for these
products, include a series of cracking reactions in which
free radical formation and recombination takes place. Only
those products that are relatively stable under the temperature conditions used persisted. Thus, as a temperature of
1000°C was approach, the simple gas composition noted was
obtained. The free carbon content in the residues decreased
and virtually disappeared. Three important reactions in
gasification are:

$$C \ + \ H_2O \rightleftharpoons CO \ + \ H_2 \qquad (1)$$

$$C + CO_2 \rightleftharpoons 2\ CO \qquad (2)$$
$$C + 2\ H_2 \rightleftharpoons CH_4 \qquad (3)$$

Once formed the relative molar composition of the products of reactions (1) through (3) are governed by reactions including (4) and (5).

$$C + H_2O \rightleftharpoons CO_2 + H_2 \qquad (4)$$
$$CH_4 + H_2O \rightleftharpoons CO + 2\ H_2 \qquad (5)$$

Clearly, other reactions play significant roles in producing final products. One important reaction, in the presence of organic material containing sulfur which is reduced to sulfide with release of hydrogen sulfide, is shown in Eq. (6).

$$Na_2CO_3 + H_2S \rightleftharpoons Na_2S + CO_2 \qquad (6)$$

The role of this reaction is particularly significant in the pyrolysis-gasification combustion of kraft black liquor. Above about 750°C this reaction is driven to formation of sodium sulfide. Thus, a mechanism for absorbing the hydrogen sulfide formed is inherent in the system. Similarly, chlorides, so troublesome in direct combustion of fuels containing chlorine (e.g., polyvinyl chlorides, etc.) may be neatly controlled in the PGC system in the presence of alkaline carbonates.

Equilibrium conditions will be approached in a flow-through reactor, such as the PGC experimental unit we have designed and operated, only in those reactions which occur rapidly. Clearly kinetics, which are not considered here, may play a controlling role in determining the composition of gaseous products formed in a reactor operating under steady state conditions. Reactor design can be modified, within limits, to favor formation and islation of desired products that can be controlled by kinetics.

It is frequently stated that a pyrolysis procedure cannot compete with direct combustion procedures in efficient recovery of energy. This paper, using a procedure recently described [8, 9] makes such a comparison on the basis of energy recovery. Clearly, a complete comparison must be made including economics, safety, environmental impact, and other considerations. Except for qualitative comments, such comparisons are beyond the scope of this paper. Based on reasonable assumptions, the same conditions when possible, and rate rather than capacity, it is shown that a pyrolysis-gasification-combustion system may be more energetically efficient than direct combustion.

In previous work [10, 11] pyrolysis-gasification was carried out at 1000°C under conditions approaching complete gasification of the wood used as a fuel stock. The data obtained has been used to calculate the equilibrium constant of reaction (4). The small residue of carbon and condensible organic products isolated were calculated as products in the ratios present in the gas phase isolated. Then, k at 1273°K (1000°C) $= \dfrac{(H_2)(CO_2)}{(CO)(H_2O)} = 0.65$. This is in good agreement with literature values which are more nearly approached than the experimental value obtained using kraft black liquor. Experimentally it was shown that k at 1273°K (1000°C) = 0.45.

In this analysis it was necessary to compare two systems, not just the energy produced in two boilers. Since utilization of energy by direct combustion necessarily requires the production of steam, this is a primary condition stipulated. Secondly, though a variety of raw materials can be considered, the feed to be considered for present purposes is wood having a specified composition. Thirdly, and most significantly, it was necessary to select a given moisture content of the wood. Drying organic matter presents an environmental problem that requires specific attention. Volatile organic materials are evolved in high yields, and must be eliminated by processing if drying is to be considered a viable operation for use in the future. The process comparison to be made should include a means of recovery or destruction of such material. To avoid this very real problem the fuel in this comparison is used in the green condition and flow diagrams were developed on this basis. Moreover, emissions of particulate material must be reduced to acceptable levels. This is provided for in the two systems. It was further assumed that the processes would be of a capacity to utilize efficient turbogenerators.

IV. SYSTEM COMPARISON

A.   The Pyrolysis-Gasification-Combustion System

   A flow schematic for a pyrolysis-gasification-combustion system is presented in Figure 1. Enthalpies of material flow streams are designed (H) and heat losses from major equipment items are designated (Q). These are used in the energy balances given.

1. First stage reactor

   Green wood in a rough "as delivered" condition, with minimal refining and separation, is introduced to the first stage reactor. In this reactor the feed is brought to a temperature that would provide for evaporation of moisture and heating of the wood-vapor stream such that pyrolysis with formation of tars will not take place. As the exit temperature of the material flow from this reactor is increased, the efficiency of the process will be increased. Conservatively, a temperature of 300°C has been used. Indirect heating is required in this reactor and is effected using the stream of gas from the waste heat boiler.

2. Pyrolysis gasification reactor

   Wood and vapors introduced to this reactor are brought to reaction temperature using a stream of preheated air. Oxygen or oxygen enrichment of air would increase the efficiency of the process. Use of oxygen is a decision to be made on the basis of economics. The reactor lining would be constructed of ceramic material capable of withstanding operating temperature of 1000°C or higher. Solid ash is removed from the reactor and pyrolysis-gasification gases (PG gas) introduced to a waste heat boiler.

3. Waste heat boiler

   Gases introduced to the waste heat boiler from the PG reactor may require cooling to an appropriate temperature by dilution with PG gas recycled from an appropriate point between the first stage reactor and gas storage. The heat removed in the waste heat boiler is adjusted so that

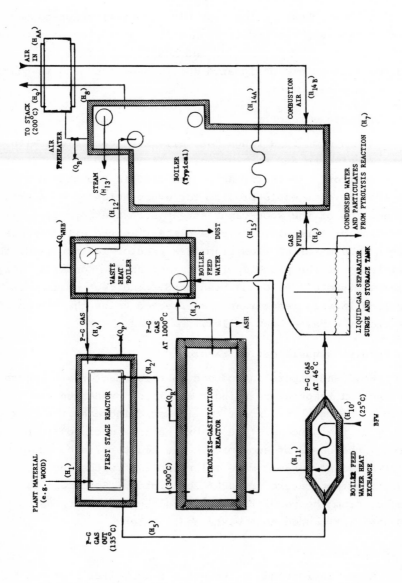

Fig. 1. Pyrolysis-gasification-combustion of plant material.

106

preheating of the wood flow through the first stage and of
boiler fuel water can be optimized. The waste heat boiler
is used to heat the steam introduced to the PG gas boiler.

The stream of PG gas from the waste heat boiler is
used to heat the first stage reactor. Temperature of exit
gases from the first stage reactor "jacket" is assumed to be
135°C.

4. Boiler feed water heat exchanger

The PG gas introduced to the boiler feed water heat
exchanger contains particulate material. The temperature of
the PG gas stream leaving this unit is reduced to the lowest
practical level; 46°C (115°F) was assumed. Under these conditions water vapor in the PG gas is condensed. Thus, this
heat exchanger is designed as a scrubber-condenser. It is
assumed that particulate material in the PG gas at this point
is effectively removed by nucleation and removed with water
condensate from gas storage. The level of particulate
remaining in the PG gas leaving gas storage is within speci-
fications. Thus, further gas treatment by electrostatic
precipitation or other techniques is unnecessary.

5. Liquid-gas separator, surge and storage tank

This tank separates condensate from the scrubbed PG
gas. It also stores a limited supply of PG gas and accommo-
dates surges in gas produced. Since all organic matter is
converted to CO, CO_2, and H_2O in the pyrolysis-gasification
reactor the condensate would be free of organic contamin-
ants and would contain only dust removed from the PG gas
stream.

6. PG gas boiler

The gas stream delivered as a fuel to the PG gas
boiler is clean; thus, the efficiency attainable in a typical
gas boiler should be realized. Moreover, the boiler feed
water stream has been preheated to a high temperature. The
clean combustion gases would be used to preheat air used in
the boiler itself as well as air introduced into the
pyrolysis-gasification reactor. To bring the air to be used

in the PG reactor to the highest practical temperature, it
is further preheated in the PG boiler as shown.   It is
assumed that temperature gases to the combustion stack would
be reduced to 200°C.   Further, the PG gas boiler and air
preheater have been considered as a unit with internal recycle for present purposes.   Air flow and boiler feed water
flow in the system have been described.

7.   Turbine

The turbine to be used with this system is not indicated in Figure 1.   However, in view of the design utilizing
low level heat for preheating the boiler feed water and then
further heating the boiler feed water stream in the waste
heat boiler, extracting steam from the turbine to preheat
boiler feed water is not required.   Accordingly, a complete
expansion condensing type turbine can be used with the PG
boiler.

8.   Material balance

The flow of material in the pyrolysis-gasification-
combustion system is given in Table I based on 1000 kg of OD
wood.   Air required for heating the PG reactor was calculated
using the procedure previously described [9].   For present
purposes it was assumed the entire ash content of the wood is
carried as dust into the waste heat boiler where it is
removed as dust.   The dust removed in the boiler feed water
heat exchanger (scrubber-condenser) is considered trivial for
purposes of this calculation.   It is assumed that 7.5%
excess air will be required for complete combustion of the
scrubbed PG gas.

9.   Energy balance

A gross energy balance has been estimated for the pyrolysis-gasification-combustion system as shown in Table II.
Power requirements in the process are not considered.   Thus
the induced draft and forced draft fans (not shown) would
require significant amounts of energy.

The energy equation used is given for each major equipment item.   Conditions used are indicated.   Specific heats

were calculated over the temperature ranges indicated using
standard literature sources. Heat losses were assumed to be
1% of the total heat introduced into major pieces of equipment. Sensible heat and heat of combustion were presented
as separate items except as noted where summations provided
clarity.

B.   The Direct Combustion System

A flow schematic for a direct combustion system is
presented in Figure 2 with enthalpies of streams and heat
losses designated as in Figure 1. Wood drying is not used
for reasons discussed above.

1. Hogger

Wood or other plant material introduced to a high
energy boiler would necessarily have to be reduced to a
particle size that can be burned efficiently. The wood
preparation is a major step and would require appreciable
amounts of energy. This requirement is neglected in this
presentation

2. Hogged fuel boiler

The hogged fuel is introduced into the boiler using
an appropriate mechanism. In order to make a comparison, it
was assumed that the same quality of steam could be produced
in the boiler utilizing hogged fuel as in the boiler using
gas. Air would be preheated using flue gas as indicated.
Induced draft and forced draft fans have been omitted and the
energy consumed by these units has also been neglected as in
the PGC system.

3. Dust removal

A system has been provided indicating the necessity of
dust removal from the combustion gas stream before release to
the atmosphere. For this purpose cyclones and an electrostatic precipitator have been indicated.

Table I. Pyrolysis–Gasification–Combustion Material Balance

[basis = 1000 kg = 1 metric ton of wood (O.D.)]

	In	Flow	Out Product
First Stage Reactor			
Wood, Organic	997.0*	997.0	
Ash	3.0*	3.0	
Water in Wood	1,103.0*	1,103.0	
Pyrolysis–Gasification Reactor			
Wood, Solids	1,000.0		
Water	1,103.0		
Total		2,103.0	
Air			
Oxygen	494.9		
Nitrogen	1,615.1		
Argon	27.6		
Carbon dioxide	1.1		
Total		2,138.7	
Water (g)	11.6		
Total		2,150.3*	
Pyrolysis Gas			
Hydrogen		⟩ 65.1	
Carbon monoxide		502.8	
Carbon dioxide		944.0	
Methane		26.5	
Nitrogen		1,615.1	
Argon		27.6	
Total		3,181.1	
Water (g)		1,069.2	
Total		4,250.3	
Dust		3.0	
Total		4,253.3	
Waste Heat Boiler			
Dry PG gas	3,181.1	3,181.1	
Water (g)	1,069.2	1,069.2	
Total	4,250.3	4,250.3	
Ash	3.0		3.0†
Total	4,253.3		
First Stage Reactor Jacket			
Dry PG gas	3,181.1	3,181.1	
Water (g)	1,069.2	1,069.2	
Total	4,250.3	4,250.3	

Table I. Pyrolysis–Gasification–Combustion Material Balance (continued)

	In Flow kg/mt, OD wood	Out Flow kg/mt, OD wood	Product kg/mt, OD wood
Boiler Stage Reactor Jacket – Gas Storage Tank			
Dry PG gas	3,181.1	3,181.1	
Water (g)			
Vapor (g)	1,069.2	83.3	
Condensate	---	---	985.9
Total	4,250.3	3,264.4	985.9†
Pyrolysis Gas Boiler			
Dry PG gas	3,181.1		
Water (g)	83.3		
Total	3,264.4		
Air (7.5% excess) and Combustion Products			
Oxygen	977.9		68.3
Nitrogen	3,191.2		4,806.3
Carbon dioxide	2.2		1,808.9
Argon	54.4		82.0
Total	4,225.7		6,765.5
Dry gas	4,225.7		6,765.5
Water	22.9		747.4
TOTAL	4,248.6*		7,512.9†
Summation for Process	8,501.9 (Σ of *)		8,501.8 (Σ of †)

Table II.   Pyrolysis–Gasification–Combustion Energy Balance

[Datum Temperature = 25°C (77°F)]
(Refer to Figure 1 for symbols)

	M kg-cal/OD Metric Ton of Wood	
	In	Out

1. First Stage Reactor
    $(H_2 + H_5 + Q_P = H_1 + H_4)$

Wood Flow
    Heat of Combustion $\underline{1}/$ (m wood × h_c wood)

$$H_1 = 1,000 \text{ kg} \times 4,730 \frac{\text{kg-cal}}{\text{kg}} =$$

[4,730.0]

Sensible Heat = $m_i \int_{25°C}^{300°C} c_{P_i} \, dt + m_{H_2O} \, \Delta h_{fg}$

Wood; 1000 kg × 0.56 $\frac{\text{kg-cal}}{(\text{kg})(°C)}$ × 275 =

154.0

Water (g); 1,103 kg × 709.2 $\frac{\text{kg-cal}}{\text{kg}}$ =

782.3
936.3
+ [4,730.0]

$H_2 =$

5,666.3*

PG Gas Flow
    Out of Jacket (H_5)
        Dry; 3,181.1 kg @ 135°C
            C_p (avg) = 0.31436 $\frac{\text{kg-cal}}{(\text{kg})(°C)}$

            3,181.1 × 0.31436 (135°C − 25°C) =

110.0

        Water (g); 1,069.2 kg × 633.56 $\frac{\text{kg-cal}}{\text{kg}}$ =

677.4

        H_c (PG Gas) = (see Item 2, PG Reactor) =

[3,780.7]

        $H_5 =$

4,568.1*

    Into Jacket (H_4)
        $(H_4 - Q_P = H_2 + H_5 - H_1)$
            = 936.3 + [4,730.0]
            + 787.4 + [3,780.7]
            − [4,730.0]
            = 1,723.7 + [3,780.7]
        Q_P = (1,723.7 + Q_P) × 0.01

21.9

        $H_4 =$

1,741.1
+ [3,780.7]
5,521.8*

Summation: Sensible Heat

	1,741.1	1,741.1
H_c – Wood Flow	[4,730.0]	[4,730.0]
– PG Gas Flow	[3,780.7]	[3,780.7]

$\underline{1}/$ Heats of Combustion are given in [ ] except when included in summations
marked by an asterisk, *.

Table II. Pyrolysis-Gasification-Combustion Energy Balance (continued)

	M kg-cal/OD Metric Ton of Wood	
	In	Out

2. Pyrolysis-Gasification (PG) Reactor
$(H_3 + Q_R = H_2 + H_{15})$

Wood + Moisture Flow
$\quad H_2$ = (see Item 1, First Stage Reactor) =   936.3

Air Flow   +[4,730.0]
$\quad H_{15}$ = Air Preheat to 604°C

$\qquad$ 2,105.3 kg moist air ×

$\qquad$ 146 $\frac{kg\text{-}cal}{kg}$ moist air =   307.5

PG Gas Flow

$$m_i \int_{25°C}^{1000°C} c_{p_i} \, dt + m_{H_2O} \, \Delta h_{fg} + \Sigma m_i \, H_{c_i}$$

Out
$\quad$ Dry PG Gas; 3,181.1 kg @ 1000°C

$\qquad c_p$ (avg) = 0.32635 $\frac{kg\text{-}cal}{(kg)(°C)}$

$\qquad$ 3,181.1 × 0.32635 × 975 =   1,012.2

$\quad$ Water (g); 1,069.2 kg × 1,083.4 $\frac{kg\text{-}cal}{kg}$ =   1,158.4

$\quad \Sigma \, h_{c\,H_2} + h_{c\,CO} + h_{c\,CH_4}$ =   [3,780.7]

$\quad$ Dust: Assume ash in wood is carried into
$\qquad$ the waste heat boiler; assume

$\qquad K_2CO_3$ 3 kg × 0.21633 $(C_p ot)$ × 975 =   0.6

$\quad H_3$ =   5,951.9*

Q_R = (1,012.2 + 1,158.4 + 0.6 + Q_R) × .01 =   21.9

Summation: Sensible Heat   1,243.8   2,193.1

$\qquad H_c$ - Wood Flow   [4,730.0]   ---

$\qquad$ - PG Gas Flow   ---   [3,780.7]

$\qquad$ 5,973.8*   5,973.8*

Table II. Pyrolysis–Gasification–Combustion Energy Balance (continued)

	M kg-cal/OD Metric Ton of Wood	
	In	Out

3. Boiler Feed Water Heat Exchanger (BFWHE)
   $(H_6 + H_7 + H_{11} = H_5 + H_{10})$
   PG Gas Flow
   In

   H_5 = (see Item 1, First Stage Reactor) =   787.4
                                                 + [3,780.7]

   Out
   Sensible Heat
      Dry; 3,181.1 kg @ 46°C

      Cp (avg) = 0.31887 $\frac{kg\text{-}cal}{(kg)(°C)}$

      3,181.1 × 0.31887 =   21.3

      Water (g); 83.3 kg × 592.39 $\frac{kg\text{-}cal}{kg}$ =   49.3
                                                              70.6
   H_c (PG Gas) (see Item 2, PG Reactor) =   [3,780.7]

   H_6 =   3,851.3*

   Condensate; H_2O (g), 985.9 kg ×

   H_7 = 985.9 kg × 21.06 $\frac{kg\text{-}cal}{kg}$ =   20.8

   Boiler Feed Water
   In

      H_{10} @ 25°C =   0.0

   Out

      H_{11} = 787.4 + [3,780.7] + 0.0
             − 3851.3* − 20.8 =   696.0

   Summation: Sensible Heat   787.4   787.4
              H_c, PG Gas Flow   [3,780.7]   [3,780.7]

4. Waste Heat Boiler
   $(H_4 + H_{12} + Q_{WHB} = H_3 + H_{11})$
   Pyrolysis Gas Flow
      H_3 = (see Item 2, PG Reactor) =   2,171.1
                                            + [3,780.7]

      H_4 = (see Item 1, First Stage Reactor) =   1,741.1
                                                         + [3,780.7]

   BFW Flow
      H_{11} = (see Item 3, BFWHE) =   696.0
   Heat Loss
      Q_{WHB} = (2,171.1 + 696.0 + Q_{WHB}) × 0.01 =   29.0
   Steam Flow
      H_{12} = 2,171.1 + 696.0 − 1,741.1 − 29.0 =   1,097.0

   Summation: Sensible Heat   2,867.1   2,867.1
              H_c (PG Gas)   [3,780.7]   [3,780.7]

Table II. Pyrolysis–Gasification–Combustion Energy Balance (continued)

	M kg-cal/OD Metric Ton of Wood	
	In	Out

5. PG Gas Boiler + Air Preheater Unit
$(H_9 + H_{13} + H_{15} + Q_B = H_{c_{PG\ Gas}} + H_6 + H_{12})$

PG Gas @ 46°C

    H_6 = (see Item 3, BFWHE) =   70.6

    $H_{c_{PG\ Gas}} = h_{c\ H_2} + h_{c\ CO} + h_{c\ CH_4}$ =   [3,780.7]

Air @ 25°C; H_{AA} =   0.0

    H_{15} = (see Item 2, PG Reactor) =   307.5

Combustion Gas @ 200°C in Stack

    Dry; 6,765.5 kg

        C_p (avg) = 0.23924

        6,765.5 × 0.23924 (200°C − 25°C) =   283.3

    Water (g); 747.4 kg × 662.22 $\frac{kg\text{-}cal}{kg}$ =   494.9

        H_9 = 283.3 + 494.9 =   778.2

Steam

    H_{12} = (see Item 4, Waste Heat Boiler) =   1,097.0

Heat Loss

    Q_B = (3,780.7 + 70.6 + 1,097.0 + Q_B) × 0.01 =   50.0

Net Heat to Steam @ 1,300 psig, 925°F

    H_{13} = 3,780.7 + 70.6 + 1097.0

           − 778.2 − 307.5 − 50.0 =   3,812.6

Summation: Sensible Heat   1,167.6   4,948.3

         H_c (PG Gas)   [3,780.7]

                                4,948.3

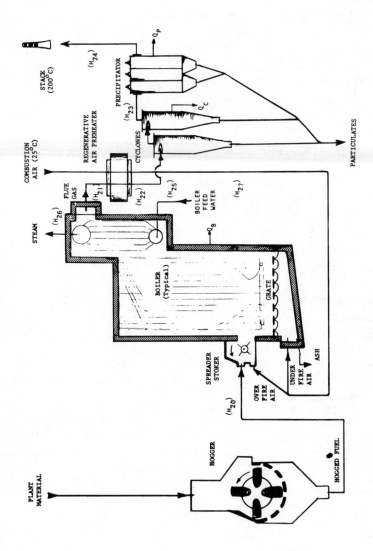

Fig. 2. Direct combustion of plant material.

116

4. Turbine

The turbine is not indicated in Figure 2. Since
boiler feed water can be reheated most efficiently by extraction of steam from the turbine, a regenerative type using
4 extraction points was assumed. Thus, a condensing regenerative turbine is indicated assuming the capacity of the
boiler requires this type.

It is also assumed that boiler feed water can be
reheated more efficiently using the turbine indicated than
using flue gases. Accordingly, the heat in the flue gas
from the boiler is assumed to be sufficient to provide the
heat required in the air preheater, and for the heat losses
from the cyclones and electrostatic precipitator and the
stack gas at 200°C.

5. Material balance and energy balance

A material balance for the direct combustion system is
given in Table III. Refuse other than ash content of wood
is assumed to be removed from the boiler, would be variable
depending on the raw material used and has been neglected
for present purposes. It is also assumed that 0.5% carbon
based on the OD wood would be lost in the refuse.

It is assumed that 37.5% excess air would be required
for complete combustion. This is an average value for a
boiler and feed of the types designated.

A gross energy, estimated for the direct combustion
system, is presented in Table IV. Conditions used are
indicated and heat losses have been accounted for as
previously described.

C. Comparative Power Generated

The net heat generated and present in the steam flowing
from the boiler is given in the energy balances for the two
cases. This heat provides the basis for comparing the relative efficiencies of the processes. However, the systems
approach is required in making this comparison because of
differences provided by the two systems. Thus, the major

Table III.   Direct Combustion Material Balance

(Basis: 1000 kg   1 Metric Ton of Wood, OD)

Material	Kg.						
INPUT	C	H	O	N	Ar	Ash	Total
Wood	492.5	65.8	438.7	--	--	3.0	1,000.0
Water	--	123.4	979.6	--	--	--	1,103.0
Air [1/]							
O_2	--	--	1,920.6	--	--	--	1,920.6
N_2	--	--	--	6,272.4	--	--	6,272.4
CO_2	1.2	--	3.1	--	--	--	4.3
Ar	--	--	--	--	107.0	--	107.0
dry							8,305.7
H_2O	--	5.0	39.9	--	--	--	44.9
Total	493.7	194.2	3,381.9	6,272.4	107.0	3.0	10,452.2
Combustion Products							
O_2	--	--	537.4	--	--	--	537.4
CO_2	488.7	--	1,303.2	--	--	--	1,791.9
H_2O	--	194.2	1.541.3	--	--	--	1,735.5
N_2	--	--	--	6,272.4	--	--	6,272.4
Ar	--	--	--	--	107.0	--	107.0
Ash	5.0	--	--	--	--	3.0	8.0
Total	493.7	194.2	3,381.9	6,272.4	107.0	3.0	10,452.2

[1/] 37.5% excess.

amount of heat used in evaporating water in the feed material
is recovered in the PGC system by utilizing the heat in the
PG gas stream to preheat boiler water. This heat is lost in
the direct combustion process. A second difference is the
possibility of using substantially reduced amounts of air
since lower excess air can be used in the PG gas boiler.
Another significant difference is the use of a complete
expansion condensing turbine in the PGC system, in contrast
to the need for a regenerative condensing type turbine for
maximum efficiency in the direct combustion procedure.

Results of the comparison are tabulated in Table V.
The PGC system, as outlined, will produce a greater amount of
electrical power than the direct combustion system. Standard
literature sources have been used for terminology and information required in the calculations made, e.g. reference [12].

V. CONCLUSIONS

It should be recognized that important considerations
have been omitted in making this comparison. These would
change the results obtained and could change conclusions to
be drawn. However, on the basis of energy alone, a decision
cannot be made concerning the selection of either process.
Indeed, a complete system analysis is required. Special
conditions or considerations prevailing in a particular
situation would have to be considered. Economics will play
an important role in such an evaluation. Economics are well
established for direct combustion, but will require estima-
tion for the PGC system. In addition, major design studies
for the PGC or other pyrolysis sytems are required for such
equipment items as the first stage reactor, the PG reactor,
the waste heat boiler, and the boiler feed water heat
exchanger. It is important that the PGC system can be controlled to provide an essentially emission free system.

Pyrolysis-gasification-combustion provides a fuel gas
that can be stored, transported, and processed. The system
can be used in parallel with other energy producing systems
and can be adapted to a wide range of capacities. For
example, it has been suggested for implemention in pulp mills
where organic residues are or can be made available for
needed energy production. It has also been suggested that
municipal solid wastes can be integrated into such energy
production.

Finally, very substantial efforts should be directed,
on a national basis, to establish the most efficient overall
systems for utilizing one of our renewable energy resources,
organic residues. Under conditions which will almost certainly prevail in future years, these materials will be an
asset and not a liability.

Table IV. Direct Combustion Energy Balance

(Datum Temperature = 25°C)
(Refer to Figure 2 for symbols)

	Mkg-cal/OD Metric Ton of Wood	
	In	Out

Overall Energy Balance
$(H_{24} + H_{26} + H_{Ash} + Q_B + Q_P + Q_C =$

$H_{20} + H_{25} + H_{AA})$

1. Wood Flow @ 25°C
   Heat of Combustion
   H_{20} = 1000 kg x 4,730 $\frac{kg\text{-}cal}{kg}$ 4,730.0

2. Boiler Feed Water (Assume preheat
   same as in PGC case; extract steam
   at required pressure and temperature from turbogenerator for BFW
   heating.)
   H_{25} =   696.0

3. Air Preheat - ambient air @ 25°C
   H_{AA} = 0; also

   $H_{22} - H_{21} = H_{27} - H_{AA};$

   assume internal recycle same
   as used in PGC case; Q_B =

   heat loss from the boiler -
   air preheater system;

4. Heat Loss in Combustion Gas
   Stack Gas Temperature 200°C

   $$H_{24} = m_i \int_{25°C}^{200°C} c_{pi}\, dt + m_{H_2O}\, \Delta h_{fg}$$

   Dry; 8,708.7 kg
   c_p (avg) = 0.23900

   = 8.708.7 x .23900 x (200°C - 25°C)=   364.2

   Water (g); 1,736 kg x 662.22 $\frac{kg\text{-}cal}{kg}$ 1,149.6

   H_{24} = 364.2 + 1,149.6 =   1,513.8

Table IV. Direct Combustion Energy Balance (continued)

	Mkg-cal/OD Metric Ton of Wood	
	In	Out

5. Heat Loss in Ash
   Ash, carbon
   $5 \text{ kg} \times 8{,}026 \frac{\text{kg-cal}}{\text{kg}} = 40.1 \text{ Mkg-cal}$
   Mineral (as K_2CO_3)(at $1000^{\circ}C$)
   $3 \text{ kg} \times 0.21633 \frac{\text{kg-cal}}{(\text{kg})(^{\circ}C)} \times 975 = 0.6 \text{ Mkg-cal}$
   0.6
   $H_{Ash} = 40.1 + 0.6 =$ 40.7

6. Heat Loss from Boiler – Air Preheater Unit
   $Q_B = (4{,}730.0 + 696.0) \times .01 =$ 54.3

7. Heat Loss from Precipitator
   $H_{23} = 1{,}513.8 + Q_P$
   $Q_P = H_{23} \times 0.01$
   $H_{23} = 1{,}529.1 \frac{\text{Mkg-cal}}{\text{mt(OD,wood)}}$
   $Q_P =$ 15.3

8. Heat Loss from Cyclones
   $H_{22} = 1{,}529.1 + Q_C$
   $Q_C = H_{22} \times 0.01$
   $H_{22} = 1{,}544.5 \frac{\text{Mkg-cal}}{\text{mt(OD,wood)}}$
   $Q_C =$ 15.4

9. Net Heat to Steam at 1300 psig, $925^{\circ}F$
   $H_{26} = 4{,}730.0 + 696.0 - 1{,}513.8 - 40.7$
   $\div 54.3 - 15.3 - 15.4 =$ 3,786.5

Table V. Comparative Power Generated

System[1]	PGC	DCB
Net Heat, Boiler = Heat in entering turbine flow		
Mkg-cal/metric ton, OD wood	3,812.6	3,786.5
MBtu/metric ton, OD wood	15,129.6	15,026.0
Turbine type	condensing complete expansion	condensing regenerative (4 heaters)
Steam Flow = W_1[2]		
kg/(hr)(mt, OD wood)	4,887.6	4,847.2
lb/(hr)(mt, OD wood)	10,775.3	10,686.3
Feedwater[3] - final heater		
h_8 = kg-cal/kg	–	168.48
= Btu/lb	–	303.27
HR_{nx}[4], kg-cal/kw hr.	5,919	5,919
Btu/kw hr.	10,654	10,654
Power generated (Pg), kw/mt, OD wood =	1,420[5]	1,308[6]
Process Efficiency - Overall Conversion to electrical power (basis Btu in wood), %	25.8	23.8

[1] PGC = Pyrolysis-Gasification-Combustion
DCB = Direct Combustion Boiler

[2] Conditions:
Assume 1300 psig, 925°F,
h_1 = 1451.2 Btu/lb of steam
s_1 = 1.5860

[3] Preheat, based on PGC system, H_{11}:

$$\Delta h_8 = \frac{696 \text{ Mkg-cal/(mt)(hr)} \times 3.96832 \frac{\text{Btu}}{\text{kg-cal}}}{10,686.3 \text{ lb/(hr)(mt)}}$$

$$= 258.26 \frac{\text{Btu}}{\text{lb}} \text{ of BFW}$$

$h_8 = \Delta h_8 + h_f = 258.26 + 45.01 = 303.27$

[4] Final exhaust, assume 1 in. Hg.
h_2 = 851.5 Btu/lb
Available Energy = $h_1 - h_2$ = 599.7 Btu/lb of steam
Turbine efficiency (both systems, assume 75%)

4/ (continued)

Actual Steam Rate = $\dfrac{3412.75}{599.7 \times .75}$ = 7.588

Nonextraction Heat Rate = HR_{hx} = 7.588 (1451.2-47.1) = 10,654 $\dfrac{Btu}{kw\ hr}$

h_{fx} = Saturated liquid at exhaust pressure = 47.1 $\dfrac{Btu}{lb.}$

5/ P_G (PGC) = $\dfrac{W_1\ (h_1 - h_{fx})}{HR_{nx}}$

(assume no steam leaks)

$= \dfrac{10,775.3\ (1451.2-47.1)}{7.558\ (1451.2-47.1)}$ = 1420 kw

6/

P_G (DCB) $= \dfrac{W_1\ (h_1 - h_8)}{Extraction\ HR}$

Gain due to 4 heaters = 10.2%

Cycle loss   _ 0.5%

    Net gain   = 9.7%

10.654 x (1-0.097)   = 9621 $\dfrac{Btu}{kw\ hr}$

Assume decrease in exhaust loss = 2.5%

9621 x (1-0.025) = 9380 $\dfrac{Btu}{kw\ hr.}$

Extraction Heat Rate = 9380 $\dfrac{Btu}{kw\ hr.}$

then

P_G (DCB) $= \dfrac{10,686.3\ (1451.2-303.3)}{9380}$ = 1308 kw

REFERENCES

1. L. L. Anderson, "Energy Potential from Organic Wastes:
    A Review of the Quantities and Sources," U. S. Dept.
    of Interior, Bureau of Mines, Information Cir. 8549,
    (1972).

2. C. G. Golueke and P. H. McGaukey, "Comprehensive Studies
    of Solid Wastes Management," Sanitary Engineering
    Research Laboratory Report, Univ. of California,
    Berkeley, June 1970.

3. D. L. Brink, J. Appl. Polym. Symp. 28, (1976).

4. Solar Energy Definition Report, Energy Research and
    Development Administration (1975).

5. Project Independence Report, Energy Research and
    Development Administration (1975).

6. "Corrosion Studies in Municipal Incinerators," Environ-
    mental Protection Agency Solid Waste Report Sw/72-3-3.

7. J. B. Grantham, Chem 26, Paper Processing 10(2), 17
    (1974).

8. D. L. Brink, G. W. Faltico, and J. F. Thomas, "The
    Pyrolysis-Gasification-Combustion Process." Presented
    to the Institute of Paper Chemistry-Pulp and Paper
    Research Institute of Canada Forum on Kraft Recovery
    Alternatives, Appleton, Wisconsin, April 30, 1976
    (in press)

9. D. L. Brink, G. W. Faltico, and J. F. Thomas, "Pyrolysis-
    Gasification-Combustion. Feasibility in Pulping
    Recovery Systems -- The First Stage-Second Stage
    Reactor as a Production Unit." TAPPI Conference
    Papers: Alkaline Pulping and Testing, Dallas, Texas,
    Sept. 13-15, 1976.

10. R. T. Williams and D. L. Brink, "In Comprehensive Stud-
    ies of Solid Wastes Management." Final Report by
    J. M. McFarland, et. al. SERL Report No. 72-3,
    Sanitary Engineering Research Laboratory, Univ. of
    California, Berkeley, 107-138, May 1972.

11. _____, "Pyrolysis-Combustion of Wood and Solid Waste Materials," MS Thesis, Univ. of California, Berkeley June 1971.

12. J. K. Salisbury, Kent's Mechanical Engineers' Handbook, Power, 1950.

CATALYTIC CONVERSION OF CARBOHYDRATES TO SYNTHESIS GAS

Herbert R. Appell and Peter Pantages

Pittsburgh Energy Research Center

U. S. Energy Research and Development Administration

Pittsburgh, Pennsylvania

I. INTRODUCTION

Although the pyrolysis of carbohydrates has been investigated extensively, the catalytic decomposition of carbohydrates has received scant attention. Most of the effort in the use of catalytically active agents has been in investigations of retarding the combustion of cellulosic materials by promoting the dehydration reaction and thus minimizing the formation of combustible gases. Stern and Heinemann [1] have reported on the conversion of sawdust to synthesis gas over steel wood and alundum at 1000°C. In view of the high temperature employed, this decomposition appears to be more pyrolytic than catalytic. Halligen [2] has reported on the conversion of bovine wastes to synthesis gas, but the reactions involved partial oxidation in the presence of steam, and although catalytic agents may have been present, the reactions occurring are normally classified as thermal rather than catalytic. The closest approach to a catalytic process for gasifying wastes was reported by Hoffman and Cox [3], who used potassium carbonate in conjunction with methanation catalysts at temperatures of 600°~700°C to obtain high-BTU gas.

The purpose of the experimental work described here was to evaluate catalysts and systems for the conversion of carbohydrates to synthesis gas. In order to show significant differences in catalytic behavior, temperatures were used where the contribution of thermal reaction was small. The emphasis in the experimental program was on comparison rather than optimization.

Cellulosic wastes are believed to be a preferred raw material for a synthesis gas feedstock. Not only is this material widely available, often as an item that requires disposal, but the decomposition reaction (to synthesis gas) requires very little heat input compared to the high endothermicity of steam-carbon gasification. The methanation reaction, which also occurs, also helps to provide heat input to the system.

II. EQUIPMENT AND PROCEDURES

The experimental work, unless designated otherwise, was conducted by dropping the reactants onto a bed of catalyst or ceramic Berl saddles centered in a Vycor tube and heated in a vertical furnace. Addition of reactants was slow, about one drop (or its equivalent in solid form) every few seconds, to avoid cooling of the reaction zone. The temperature was measured by a thermocouple cemented to the outer wall of the reactor at the top of the catalyst bed. The catalyst, or inert, bed was 1 in deep. Aqueous solutions were added from a burette. Sawdust (10-40 mesh) was added from a wide-stem funnel connected to the reactor by a short length of heavy-walled rubber tubing. The feed of solids was controlled by opening and closing a pinch clamp on the tubing. The exit gases were passed through a liquid receiver in an ice bath and then collected and measured in a gas bottle.

The system was flushed with nitrogen prior to adding the reactants, and the gas composition was calculated on a nitrogen free basis.

III. RESULTS

A.   Indirect Decomposition

In the indirect conversion of carbohydrates to synthesis gas, the initial reaction of the carbohydrate is largely a dehydration to carbon plus water followed by the steam-carbon reaction to yield synthesis gas. Either of these reactions may be catalytic or noncatalytic.

Table I shows the extent of decomposition obtained by adding a glucose solution, dropwise, to a bed of potential catalysts. The first material tested, carborundum, a low-surface area inert material, serves as the equivalent of a blank run. The SK-500, a highly acid Y-type molecular sieve catalyst, gave yields higher than the blank, showing that a strong acid catalyst, although poor, is better than none. Because of the small pore structure of this catalyst, it appears that the decomposition products of glucose are reacting on the SK-500 surface to yield the gaseous products. An acidic catalyst reacting upon the glucose would be expected to be less selective, i.e., promote the dehydration reaction and yield less gas than inert material.

TABLE I.

Gasification of Glucose over Solid Catalyst[a]

Catalyst	Temp (°C)	ml gas/gm glucose	Gas composition (%)				Evolved as gas	
			H_2	CH_4	CO	CO_2	H(%)	C(%)
Carborundum	600	129	13	8	63	11	6	15
SK-500 molecular sieve	600	219	26	7	57	9	11	22
Soda lime	600	424	73	11	8	4	68	21
$SrCO_3$	600	522	21	8	30	32	29	52
$SrCO_3$	675	744	35	10	39	14	56	64
Soda lime	675	626	76	13	4	5	89	19
Soda lime[b]	675	514	59	11	12	16	57	27

a. 45.5% aqueous glucose.
b. Treated with excess CO_2 before use.

As expected, the alkaline catalysts gave the highest yields of gas. Strontium carbonate was very active at 675°C. Soda lime was unusual in that it appeared to promote hydrogen formation. This presumably occurred because of the water-gas shift reaction. This shift reaction is indicated because of the low carbon monoxide content of the final gas. Normally a high carbon monoxide content would be expected, but the soda lime may have absorbed some carbon dioxide, even though the catalyst was pretreated with carbon dioxide in one test.

Several additional tests of similar nature were conducted over charcoal catalysts (Table II). The untreated charcoal gave a higher gas yield than the acid-washed material, indicating that part of the activity was due to alkaline impurities in the original charcoal. A significant improvement in activity was obtained by adding alkaline carbonates, especially potassium, to the charcoal. Because of the small pore structure of charcoal, it is most likely that volatile decomposition products of glucose are being reformed to synthesis gas.

TABLE II.

Gasification of Glucose over Activated Coconut Charcoal[a]

Charcoal treatment	ml gas/gm glucose	Gas composition (%)				Evolved as gas	
		H_2	CH_4	CO	CO_2	H (%)	C (%)
None	351	23	13	34	26	25	21
Acid-washed	312	14	12	51	17	19	35
10% Na_2CO_3	378	43	6	9	40	30	29
10% $Ba(OH)_2$	446	44	7	12	35	37	33
10% K_2CO_3	538	56	7	7	28	52	32

a. 45.5% aqueous glucose, 600°C.

Because of the problem of contacting glucose with a catalytic surface, a series of experiments was conducted using catalysts that were dissolved or suspended in the aqueous glucose and the composite then dropped onto an inert surface in the heat reactor.

Table III lists the gas yields, gas compositions, and the percentage of the available hydrogen and carbon that

were gasified. Two types of catalysts were used. The highest gas yields were obtained with the alkaline catalysts,
especially potassium carbonate. The nickel and iron form-
ates were tested in the hope that an active metal (via
decomposition) would catalyze the direct decomposition of the
carbohydrate molecule. This was realized to some extent with
the nickel formate, but apparently, in the presence of steam
at 675°C, the iron formed was rapidly converted to iron oxide
and the catalytic activity was poor. The molybdic oxide
showed sufficiently good activity to suggest additional
research as to its mode of action and possible improvement.

TABLE III.

Gasification of Glucose with Dissolved or Suspended Catalysts[a]

Catalyst (%)[b]	ml gas/gm glucose	Gas composition (%)				Evolved as gas	
		H_2	CH_4	CO	CO_2	H(%)	C(%)
$SrCO_3$ (1)	532	33	10	41	14	39	47
$Ca(OH)_2$ (1)	589	37	10	35	16	47	49
Rb_2CO_3 (1)	337	33	12	42	19	23	34
MoO_3 (1)	647	52	6	12	29	56	41
$Ni(O_2CH)_2$, (0.5)	505	44	6	33	15	40	37
$Fe(O_2CH)_3$, (0.67)	292	39	10	20	28	24	23
K_2CO_3, (10)	762	50	7	15	25	62	50
$K_2CO_3 \cdot KCl$ (10)	735	41	7	19	26	62	52
$K_2CO_3 \cdot Na_2CO_3$ (10)	691	43	8	19	28	57	57

a. 45.5% aqueous glucose, 675°C.
b. Weight percent in the solution.

The effect of the concentration of potassium carbonate
on the gas yield and composition is shown in Table IV. The
data show considerable deviation but do indicate that small
concentrations of potassium carbonate are adequate for catalytic activity and that the concentration used in Table III,
where various catalysts are compared, is near the optimum.
The major effect of increasing the potassium carbonate concentration is to increase the water-gas shift reaction. This
is shown by the upward trend of carbon dioxide and the downward trend of carbon monoxide.

TABLE IV.

Gasification of Glucose in Potassium Carbonate Solution[a]

K_2CO_3 (wt %)	ml gas/gm glucose	Gas composition (%)				Evolved as gas	
		H_2	CH_4	CO	CO_2	H(%)	C(%)
None	243	22	8	37	28	13	25
1	299	22	10	32	32	20	33
2	336	25	9	25	37	21	33
5	267	20	9	19	48	16	29
10	371	36	8	14	39	27	31
20	332	27	10	12	45	24	32

a. 45.5% aqueous glucose, 600°C; 45.5 gm glucose, 10 gm K_2CO_3, 90 gm H_2O.
b. The weight of the element in the gaseous products divided by the weight of the element in the glucose charged.

The gas yield and hydrogen content of the product gas both increase with temperature (Table V). The gas yield more than doubles in going from 600°C to 675°C.

TABLE V.

Effect of Temperature on Glucose Gasification with K_2CO_3 [a]

Temp. (°C)	ml gas/gm glucose	Gas composition (%)				Evolved as gas	
		H_2	CH_4	CO	CO_2	H(%)	C(%)
550	281	24	9	16	46	19	28
600	371	36	8	14	39	27	31
625	535	37	9	15	39	39	48
675	762	50	7	15	25	62	50

a. 45.5% glucose in 10% aqueous K_2CO_3.

In all of the experiments using alkaline catalysts, the gas evolution (following the addition of a drop of solution) was rapid for the first few seconds and then continued to evolve gas at a decreasing rate for another 5 to 10 seconds. This slow rate of gas evolution is believed to be a result of the steam-carbon reaction with increasing refractory carbon.

It appeared likely that if the free radical decomposition of the carbohydrates could be promoted, less dehydration
would occur and higher reaction rates and selectivities could
be obtained. With this objective in mind, several transition
metals known to have catalytic activity in breaking carbon-
carbon and carbon-hydrogen bonds were evaluated.

B. Direct Decomposition

The direct approach to the gasification of aqueous
glucose was catalysis by massive metal (Table VI). Nichrome
showed much better catalytic activity than 316 stainless
steel, but Monel alloy and nickel gave the highest results.
The importance of surface was indicated by the unusually high
gas yield with the small nickel turnings. The influence of
the metal is much greater than the amount of surface available.
The stainless steel and nichrome were in the form of a fine
wire gauge with considerably more surface than the nickel
turnings.

TABLE VI.

Gasification of Glucose over Metals[a]

Metal	Ni(%)	ml gas/gm glucose	Gas composition (%)				Evolved as gas	
			H_2	CH_4	CO	CO_2	H(%)	C(%)
Stainless steel (316)	10-14	275	36	6	35	19	19	23
Nichrome	75	495	26	9	51	9	30	47
Monel	60-70	990	48	3	43	6	71	69
Nickel shot	100	739	44	4	45	7	51	56
Nickel turnings								
large 3/8"	100	685	41	4	47	6	46	54
small 1/8"	100	1164	54	2	27	17	90	70

a. 45.5% aqueous glucose, 600°C.

A brief look at possible poisoning agents for gasification by nickel showed that the three additives tested--
sulfur, base, and acid--all caused a reduction in catalytic
activity (Table VII). The potassium carbonate probably
caused deactivation by coating the active nickel surface with
a less active catalyst. The HCl probably promoted the

glucose and thereby decreased the amount available for
direct decomposition.

TABLE VII.

Effect of Poisons on Gasification of Glucose with Nickel Turnings[a]

Poison (wt %) (in glucose)	ml gas/gm glucose	Gas composition (%)				Evolved as gas	
		H_2	CH_4	CO	CO_2	H(%)	C(%)
None	1164	54	2	27	17	90	70
$(NH_4)_2S_x$ 0.5	632	27	8	48	13	42	62
K_2CO_3 0.1	634	45	4	25	24	48	47
HCl 0.1	838	41	6	33	17	61	65

a. 45.5% aqueous glucose, 600°C.

One of the problems confronting the use of massive
metals as catalyst for carbohydrate decomposition was the
occurrence of charring before all of the reactant had an
opportunity to contact the catalyst. One approach to this
problem is the conversion of the carbohydrate to a more stable, preferably liquid, derivative. The decomposition of
sorbitol (Table VIII) in the presence of nickel shot was
conducted to evaluate this approach. Good yields of synthesis gas were obtained by dropping the sorbitol onto the catalyst in a Monel flask at temperatures of 475-525°C. The
decomposition was relatively insensitive to temperature in
this range. Although the decomposition to synthesis gas was
selective, a considerable amount of carbonaceous residue was
formed. Although the agitation in the stirred catalyst bed
kept an active surface available, the sorbitol's tendency to
carbonize resulted in relatively poor selectivity as measured by the presence of hydrogen and carbon evolved. The use
of lower temperatures should be explored.

The effectiveness of a number of metals for the direct
decomposition of cellulose was determined by impregnating
softwood sawdust with compounds that are converted to the
metal at the operating conditions of the experiment. Table
IX shows that cobalt, added as cobalt carbonyl (via impreg-
nation in pentane solution and drying), is the most active

TABLE VIII.

Gasification of Sorbitol in a Stirred Reactor[a]

Temp. (°C)	ml gas/gm sorbitol	Gas composition (%)				Evolved as gas	
		H_2	CH_4	CO	CO_2	H(%)	C(%)
475	901	38	3	43	4	53	61
500	875	46	3	45	3	61	60
525	894	39	6	48	4	53	66

a. 500 ml Monel flask, stirrer, 17 ml nickel shot.

and selective of the catalysts evaluated. One possible
explanation for the high performance of this catalyst is that
the cobalt carbonyl decomposes to a metal carbonyl at lower
temperatures than the other catalyst precursors, and a catalytic surface is formed in contact with the sawdust before
thermal dehydration can occur. Table X shows that, as the
cobalt concentration increases, higher conversions of sawdust
to gas are obtained.

TABLE IX.

Gasification of Sawdust with Impregnated Metals at 625°C

Metal (%)	Impregnating agent	ml gas/gm wood	Gas composition (%)				Evolved as gas	
			H_2	CH_4	CO	CO_2	H(%)	C(%)
None	-	593	18	16	48	12	36	44
Ag (0.25)	$AgNO_3$	594	22	15	40	16	41	42
Pd (0.008)	$PdCl_2$	662	17	16	49	10	41	49
Pd (0.25)	$PdCl_2$	684	17	16	51	9	42	52
Pt (0.12)	K_2PtCl_6	724	19	16	47	11	45	54
Co (0.25	$Co_2(CO)_8$	888	333	11	41	11	58	55

TABLE X.

Effect of Cobalt Concentration on Wood Gasification[a]

Cobalt (%)	mg gas/gm wood	Gas composition (%)				Evolved as gas	
		H_2	CH_4	CO	CO_2	H(%)	C(%)
5.1	1513	41	4	42	12	97	92
2.5	1716[b]	38	5	30	25	102[b]	103[b]
1.1	1170	42	8	39	10	84	68
0.52	896	32	10	39	14	59	56
0.25	888	33	11	41	11	58	55
None	593	18	16	48	12	36	44

a. Softwood sawdust impregnated with $Co_2(CO)_8$, 625°C.
b. Residue from previous run caused high gas yield.

Temperature also has a beneficial effect on the direct catalytic gasification with cobalt. Table XI illustrates the increase in gas yield with temperature in the 500-650°C range for pure cellulose powder and also for sawdust. The gas yields are probably higher for cellulose because this material is theoretically cleavable in high yield to carbon monoxide and hydrogen, whereas the sawdust also contains considerable lignin, which would be expected to leave some slow-reacting carbonaceous residue.

TABLE XI.

Effect of Temperature on Gasification with 0.25% Cobalt

Feedstock	Temp. (°C)	ml gas/gm feed	Gas composition (%)				Evolved as gas	
			H_2	CH_4	CO	CO_2	H(%)	C(%)
Cellulose	550	422	21	8	54	10	23	32
	575	615	28	8	51	8	33	42
	600	897	34	7	48	7	51	55
	625	919	34	7	48	7	56	59
	650	1128	39	7	45	6	72	68
Sawdust	550	534	31	10	33	20	34	34
	575	703	39	9	32	18	49	41
	600	775	39	8	33	15	49	44
	625	841	40	9	32	15	60	50
	650	973	43	8	36	11	73	54

Table XII compares a number of potential feedstocks for gasification. The relative ease of gasification was the same both in the absence and presence of a catalyst, but the gas yield was greater in the presence of cobalt. The gas obtained was also richer in hydrogen when cobalt was present, except for manure, where the presence of nitrogen and sulfur may have resulted in partially poisoning the catalyst. The lower gas yield with manure is also probably a result of catalyst poisoning.

TABLE XII.

Comparison of Feedstocks for Gasification with Cobalt at 625°C

Feedstock	Catalyst	ml gas/gm feed	Gas composition (%)				Evolved as gas	
			H_2	CH_4	CO	CO_2	H(%)	C(%)
Cellulose powder	None	764	--	--	--	--	N.D.	N.D.
Sawdust	None	593	18	16	48	12	36	44
Corncobs	None	448	14	15	44	19	29	39
Manure	None	340	22	17	29	22	24	25
Cellulose powder	0.25% Co	919	34	7	48	7	56	59
Sawdust	0.25% Co	888	33	11	41	11	58	55
Corncobs	0.25% Co	526	24	13	36	21	38	41

The uncatalyzed manure gasification also results in lower gas yields than are obtained with the other feedstocks. This is an indication of the refractory nature of this material and is most likely due to its high content of noncarbohydrate material, such as lignin.

The rapidity of all of the decompositions with transition metals indicated a direct decomposition of the organic molecule. The evolution of gas occurred rapidly and was essentially complete within a second or two after being dropped into the reaction zone. There was little evidence of the slower steam-carbon reaction occurring even in the experiments showing poor conversion to gas. It appears possible that the use of both the transition metal and the alkali

carbonate would improve the conversions obtained with the
more refractory feedstocks.

Nickel is usually a much more active catalyst than
cobalt in most reactions, including decomposition and dehydro-
genation. The high conversion of carbohydrates obtained with
nickel turnings suggested that finely divided nickel, such as
would be obtained by the decomposition of nickel carbonyl,
would be a very potent catalyst for gasification. Another
advantage of a nickel catalyst would be its ease of recovery
as the carbonyl from the residue and then reuse to impreg-
nate incoming cellulosic feedstock for gasification. Because
of the toxicity of nickel carbonyl and the precautions needed
in its preparation and use, this material was not evaluated.

C. Mechanism

Mechanistically, the direct decomposition of carbohydrates to synthesis gas probably proceeds via abstraction of
hydrogen followed by disruption of the unstable free radical
along classic lines. This type of mechanism can be expected
to operate during flash pyrolysis of carbohydrates and at
lower temperatures in the presence of transition metal catalysts that can abstract hydrogen from the substrate
(hydrogenation-dehydrogenation catalysts).

No experiments were conducted to verify the mechanism,
but it can be hypothesized to proceed by cleavage of bonds
beta to the free radical sites. The following example of a
fragment of a carbohydrate molecule with a hydrogen atom
being abstracted, and the resultant weakened bonds, shows a
possible start of the fragmentation to carbon monoxide and
hydrogen. The resulting fragments would be expected to be
highly unstable and help to account for the observed high
reaction rate.

```
    H    H    H    H              H    H    H    H
    O    O    O    O    Ni        O    O    O   | O
  - C -  C -  C -  C -  ------>  - C -  C -  C - | C -     CO + H2
    H    H    H    H              H    .    H   | H
                                      .
                                      H
```

III. CONCLUSIONS

Carbohydrates can be converted to synthesis gas by alkali metal carbonates or by transition metal catalysts. The metals cobalt and nickel show high activity and selectivity in promoting a direct decomposition of carbohydrates to hydrogen and carbon monoxide plus smaller quantities of methane and carbon dioxide.

REFERENCES

1. E. W. Stern, A. S. Logiudice, and H. Heinemann, Ind. Eng. Chem., Proc. Des. Dev. 4 (4), 171 (1905).
2. J. E. Halligan, K. L. Herzog, and H. W. Parker, Ind. Eng. Chem., Proc. Des. and Dev. 14 (1), 64 (1975).
3a. E. J. Hoffman, Preprints, Div. of Petrol. Chem., Am. Chem. Soc. 16 (2), C20 (1971).
 b. J. L. Cox, E. J. Hoffman, R. W. Hoffman, W. G. Willson, J. A. Roberts, and D. L. Stinsin, Preprints, Div. Fuel Chem., Am. Chem. Soc. 18 (1), 1-14 (1973).

MIXING URBAN WASTE AND WOOD WASTE FOR GASIFICATION

IN A PUROX REACTOR

David A. Tillman

Materials Associates, Inc.

Washington, D. C.

I. INTRODUCTION

Numerous types of organic waste are generated every day. These can and do make a significant contribution to the United States' energy supply. In the years to come, however, they will be called upon to supply increasing amounts of fuel as domestic sources of oil and natural gas continue to decline, as uranium supply becomes more limited, and as the coal-mining industry struggles to keep up to rising demand. Optimization of energy from organic residues necessarily implies investigation of innovative approaches and technologies that are capable of supplying clean, convenient fuel to markets that can utilize the BTUs made available.

To date, the programs that convert organic waste to energy treat each type of residue as an independent entity.

As a result, many concentrations are deemed too small to support waste-to-energy systems. Some sources suggest, for example, that fully half of the urban waste generated in the United States cannot be reused because concentrations of these residues are insufficient to support capital intensive resource recovery systems. If, however, wastes from dissimilar sources are combined, this problem may be overcome. Thus, the mixing of dissimilar wastes deserves serious consideration.

The volume of organic waste produced in the United States is presented in Table I. It includes two forms of waste already supplying energy to the U.S. economy--municipal solid waste (MSW) and wood waste from logging and product manufacturing operations. Those sources of organic residues already contribute, annually, over 1×10^{15} BTUs to the economy. Some 14 plants convert 3 million tons of MSW into 2.2×10^{13} BTUs. Wood waste supplies 1.1×10^{15} BTUs [1]. More plants are being committed rapidly. Other sources of

TABLE I

Estimates of Available Organic Wastes

Source	Total organic wastes generated	Organic solids available
	(million tons/year)	
Manure	200	26.0
Urban refuse	129	71.0
Logging and wood manufacturing residues	55	5.0
Agriculture crops and food wastes	390	22.6
Industrial wastes	44	5.2
Municipal sewage solids	12	1.5
Miscellaneous organic wastes	50	5.0
Total	880	136.3
Net energy potential in quadrillion BTUs	8.8	1.36

organic waste presently produce negligible amounts of energy, although plans exist to increase their supply capability dramatically.

Even though urban and wood wastes contribute a significant amount of energy to the United States economy, much room exists for their improvement. Currently, only 2.1% of the MSW generated in the United States is being converted into energy. And, although a much larger percentage of wood residues are burned for the production of steam, there is still room for improvement. In the United States as a whole, paper industry supplies 42% of its energy requirements from residue combustion [2]. In the Northeast region, however, that industry supplies far less of its own energy requirements by using residues. Neither MSW nor wood residue can achieve their full potential of energy supply if each source of organic material is treated as a separate and independent entity. A more appropriate approach is to treat them as complementary sets in the universe of renewable energy resources and to examine the potential economic and technical results that might be gained from mixing the two waste streams in a single operation.

II.   MIXING WASTE STREAMS OFFERS ECONOMIC AND TECHNICAL
      ADVANTAGES

Cellulose is one basic constituent in both MSW and wood waste, although both residue streams contain significant quantities of noncellulosic energy-bearing organics. Numerous experiments have demonstrated that, although the specific composition of the two waste streams differ, the materials are compatible. Thus, mixing of urban and wood waste prior to their conversion into energy fuels may be utilized. It improves the economic characteristics of the disposal system while moderating some of the technical problems associated with specific processes. Both the economic and processing aspects of handling blended waste, therefore, merit consideration.

A.   Economic Considerations of Mixing Waste Streams

The economic advantages of mixing waste streams stem
from increasing the quantities of residue available for conversion to energy. Because waste-to-energy systems are
capital intensive, they are size sensitive. Most are marginally economic at 500 tons per day (TPD) of residue and
fully economic at 1000+ TPD.

Many communities and localized regions do not produce
that much waste, however. Assuming that the per capita rate
of MSW generation is 3.33 lbs/day, it takes a population of
300,000 to support a 500 TPD system. One example of this
problem is the Claremont--Newport region of New Hampshire,
where the population is approximately 80,000. That region
generates only about 150 TPD of MSW. It will have to wait
some 25 years until it can support a 500 TPD energy recovery
system--assuming that its waste generation growth rate is 5%
per year.

That same region demonstrates the problem of wood
residue disposal in the Northeast. Some seventeen sawmills,
paper mills, and wood products manufacturing plants produce
100 TPD of bark, 200 TPD of sawdust, and over 200 TPD of
shavings, ends, chips, etc. [3]. No single mill generates
enough wood waste to make residue conversion to energy economically viable, however. Some chips are sold, but the vast
majority of this waste is discarded. At the same time,
lumber-drying kilns are fired with #2 distillate oil or
propane.

The Claremont--Newport, New Hampshire region is far
from unique. The relative paucity, but presence, of both
wood waste and MSW can be found throughout the northeast--
and in other wood products oriented regions of the nation as
well. The overall dimensions of the MSW problem can be
determined from the fact that virtually all studies on energy
from waste write off the municipal waste generated in
smaller communities. This conclusion implies that 22.5 -
30 x 10^6 tons of MSW generated annually can not be converted
into 2.14 x 10^{14} BTUs. Some studies conclude that 50% of
the MSW generated annually will not be converted into energy

due to the low volumes generated daily. These studies
write off 7.13 x 10^{14} BTUs. For wood wastes, the problem of
disposal is equally significant.

Wood waste availability can be viewed best by considering the problem of bark disposal. Over 10 million tons of
bark are generated in the United States every year. This
bark contains 2 x 10^{14} BTUs. This bark residue is produced
at pulping mills, sawmills, plywood plants, and veneer
plants. According to Cliff, "Except for limited commercial
uses such as mulch and fuel, most of the bark is disposed of
in landfill, by burning, or simply allowed to accumulate.
Environmental concern over such practices emphasizes the
need for more productive uses of this material, or less
offensive means of disposal" [4].

The mixing of MSW and wood waste could provide the
quantities of residue required for implementation of waste-
to-energy systems in many situations. In the Claremont, New
Hampshire area, some 650 TPD of waste would be available,
supplying over 2 x 10^{12} BTUs/yr to the local economy. Further, waste collection companies, such as Browning-Ferris
Industries, could be used to accomplish the concentration of
feedstock material.

B.   Processing Considerations of Mixing Waste Streams

The mixing of waste streams can improve the feed
material for waste-to-energy systems. Urban waste, in an as-
received form, contains many inorganic materials that inhibit
efficient conversion to energy while causing erosion and
pollution problems. Wood waste, which is inherently superior
as a feedstock, may contain significant quantities of water,
which also inhibits efficient energy production. These
problems and the impact of mixing merit investigation.

1.   Urban waste composition

Urban waste consists of approximately 74% organic
material and 26% inorganic material, as Table II demonstrates. Further, the organics contain varying amounts of
energy, ranging from 2500 BTU/lb to 15,000 BTU/lb. Table III
presents an ultimate and proximate analysis of both

TABLE II

Estimate of the Composition of Municipal Solid Waste[a]

Waste Component	Percentage of Total
Organic materials	73.6
Paper products	42.0
Food wastes	12.0
Yard wastes	15.0
Plastics	} 1.6
Rubber	
Textiles	0.6
Wood	2.4
Inorganic materials	26.4
Ferrous metal	} 8.0
Nonferrous metal	
Glass	6.0
Rocks and dirt	11.0
Other	1.4

a. Source: National Solid Waste Management
Association.

as-received and dry MSW.   These data suggest that MSW combustion presents severe efficiency, erosion or abrasion, and
pollution problems stemming from the ash, which constitutes
almost 30% of the as-received refuse.

Because MSW contains significant quantities of metal,
glass, and stone, efficiency and tribology problems plague
energy systems.   The inorganics dilute the feedstock.   They
must be heated (and cooled), consuming sensible heat without
producing energy.   The conveying systems that move the waste
into boilers are subjected to severe erosion caused by glass
fines and grit.   Expensive specialty steels must replace mild
steels in pneumatic conveying systems, especially in elbows
and other areas where abrasions may be catastrophic.   Mini-
mization of this problem calls for either separation of the
glass and metal, or dilution of the amount of inorganic
materials in the feedstock.

TABLE III
Proximate and Ultimate Analyses of MSW

Component	Analysis (wt %)	
	As received	Dry
Proximate analysis		
Moisture	20.5	0.0
Volatile matter	40.3	49.7
Fixed carbon	9.9	12.3
Ash	29.3	38.0
Ultimate analysis		
Hydrogen	6.0	4.6
Carbon	27.5	33.9
Nitrogen	0.5	0.7
Oxygen	36.4	22.4
Sulfur	0.3	0.4
Ash	29.3	38.0
BTU/lb	4570	5645

Pollution problems stem from fly ash and residue land-
filling. Since fly-ash problems are inherent in both MSW
and wood waste, waste-to-fuel systems appear desirable.
Waste-to-fuel systems are those that convert residues into
clean liquid or gaseous fuels that can then be used in the
existing boilers and kilns of industry. Residue landfill-
ing problems such as leachate generation can be handled by
converting all organics to fuel.

2. Wood waste composition
    Wood waste, which has an inherently higher BTU
content than MSW, does not have such problems as metal or
glass. Table IV shows the ultimate and proximate analysis
of as-received and dry pine bark wood waste. It illustrates the relative absence of inorganic materials, which
plague MSW, while highlighting the one deterrent to its
use in energy production--water.

TABLE IV

Proximate and Ultimate Analysis of As-Received and Dry Wood Waste[a]

Component	Analysis (wt %)	
	As received	Dry
Proximate analysis		
Moisture	50.0	0.0
Volatile matter	41.0	81.5
Fixed carbon	8.5	17.5
Ash	0.5	1.0
Ultimate analysis		
Carbon	26.2	52.3
Hydrogen	2.9	5.8
Oxygen	19.4	38.8
Nitrogen	0.1	0.2
Sulfur	0.0	0.0
Ash	1.5	2.9
Moisture	50.0	4.9
BTU/lb	3840	8780

a. Source: M. D. Schlesinger, W. S. Sanner, and D.E. Wolfson, "Pyrolysis of Waste Materials From Urban and Rural Sources," Proc. Third Mineral Waste Utilization Symp., March 14--16, 1972, p. 426.

Freshly cut wood may contain 50% or more moisture, reducing its energy value, as Table IV shows. Large volumes of moisture inhibit the efficiency of both combustion systems and waste-to-fuel pyrolysis systems. The presence of over 60-65% moisture can make combustion so inefficient that energy output is less than economic. Tests at the Pittsburgh Energy Research Center (PERC) in a research pyrolysis reactor demonstrated similar impacts in the conversion of refuse to fuel. At pyrolysis temperatures, the water-gas shift, the steam-carbon, and the steam-hydrocarbon reactions operate very rapidly, and the presence of significant

quantities of water in the feedstock shift the equilibrium
points to the right-hand side of these equations:

$$H_2O + CO \rightleftharpoons CO_2 + H_2$$
$$H_2O + C \rightleftharpoons CO + H_2$$
$$H_2O + CH_4 \rightleftharpoons CO + 3H_2$$

The PERC tests, using waste containing 7.3 and 43.3% moisture, respectively, demonstrated that 1 ton of feed materials produced 18,470 ft^3 of gas containing 545 BTU/ft^3 when
dried, and 17,741 ft^3 of gas containing 447 BTU/ft^3 when the
higher quantity of water remained in the feed material. The
energy available in the total gas produced declined by 21%
due to the presence of the additional 36% moisture [5]. The
problem of moisture in wood waste can be minimized by
dilution or solved by predrying.

3. The impact of mixing waste streams

     Mixing MSW and wood waste appears to offer the advan-
tage of alleviating problems for both feedstocks. Municipal
refuse contains 24% inorganic material; this would be diluted
reducing certain efficiency, erosion, and pollution problems
inherent in this energy source. Wood waste contains large
volumes of water, which could be reduced by blending this
feedstock with the inherently dryer municipal waste. In
the process, a useful feedstock may be created. Table V
suggests the approximate composition that can be obtained by
a 50-50 blend of urban and wood waste, while Table VI offers
proximate and ultimate analysis of this material. These
tables demonstrate that the problems endemic to individual
feedstocks can be minimized, if not solved, by the process
of blending.

III.   GASIFICATION OF THE BLENDED FEEDSTOCK BY THE PUROX^{TM}
       PROCESS

     When dissimilar organic waste streams are mixed,
direct combustion may not be an ideal approach to energy
production, because blending introduces additional hetero-
geneity into the feed material. Conversion of the feedstock to a uniform fuel solves that problem. The conversion

TABLE V

Composition of MSW and Wood Waste Blend (Dry Weight Basis)

Waste Component	Composition (%)
Organic materials	86.8
Paper	21.0
Food wastes	6.0
Yard wastes	7.5
Plastics	
Rubber	}0.8
Textiles	0.3
Wood and wood wastes	51.2
Inorganic materials	13.2

TABLE VI

Proximate and Ultimate Analysis of Blended Waste

Component	Analysis (wt %)
Proximate analysis	
Moisture	35.4
Volatile matter	40.7
Fixed carbon	9.0
Ash	14.9
Ultimate analysis	
Carbon	27.8
Hydrogen	4.2
Oxygen	17.4
Nitrogen	0.3
Sulfur	0.1
Ash	14.9
BTU/lb	4205

process studied for blended MSW and wood waste was the
Purox^{TM} process developed by the Linde Division of Union
Carbide. It produces a highly useful fuel. It also
addresses the technical considerations of inorganic materials and water.

A. Description of the Purox^{TM} Process Applied to
   As-Received Refuse

Union Carbide Corporation developed the Purox^{TM}
pyrolysis system at its Tarrytown, New York, research facility, and successfully operated a 5 TPD reactor there. Based
upon that success, it built a 200 TPD module of a commercial
plant in South Charleston, West Virginia. This test facility
has been run on both as-received refuse and on shredded,
magnetically sorted MSW. It has the capability for handling
not only MSW, but also a variety of other feed materials,
including sewage sludge dried to 20% solids, 80% moisture.

The Purox^{TM} system is a vertical shaft furnace, not
unlike a blast furnace. Wastes are charged into the reactor
vessel at the top and enter the first of three operating
zones, the drying zone. By gravity they then move into the
reaction zone where pyrolysis occurs. The gaseous and
liquid products of this pyrolysis leave the reaction zone
and move upward, while inorganics and char continue downward
into the combustion zone. At the reactor hearth, the char
is combusted with pure oxygen and fed through the tuyeres of
the reactor at temperatures of 3000°F.

The abundance of char falling from the reaction zone
creates a reducing atmosphere at the hearth. The CO
produced during combustion under reducing conditions is
released from the hearth of the reactor into the pyrolysis
zone, moving countercurrent to the feed material to insure
maximum rates of heat transfer and gasification. Pyrolysis
occurs in the reducing atmosphere created by the CO at temperatures ranging from 1200°F to 1700°F. The rapid feed
rates combine with the reducing atmosphere and the elevated
temperatures to insure a minimum of oxidation and a maximum
of gas production (with a minimum production of oil mists).

Once the pyrolysis gas is produced, it moves upward through
the feed material in drying zone to the off-take pipe. This
countercurrent flow of pyrolysis gases again insures maximum
contact with the incoming waste for drying purposes. The
gases are produced at elevated temperatures, but leave the
reactor at 400°F. Their sensible heat is used to drive
moisture from the incoming feed material[6]. Figure 1
offers a schematic of this system.

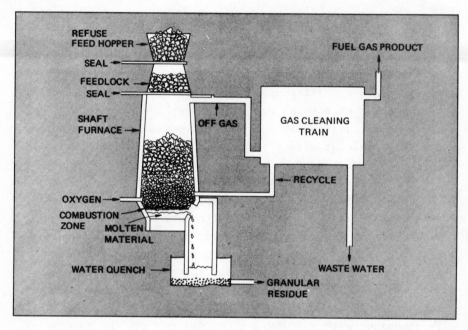

Fig. 1.   Oxygen refuse converter.

Once the gas has left the reactor, it passes through
a wet scrubber and precipitator cleaning system where oil
mist is removed and returned to the reactor for cracking
into additional gas. Fly ash is returned to the reactor
where it is destroyed. After cleaning, the dry gas is sold
to the customer. It has an average heating value of about
350 BTU ft^3, is free of potential pollutants, and can be
combusted in a manner similar to natural gas. Table VII
provides an analysis of the composition of this gas, while
Table VIII presents its combustion characteristics compared

to natural gas and propane. As these tables illustrate, the fuel gas produced in the oxygen converter are highly useful.

TABLE VII
Chemical Composition of Purox Fuel[a]

Fuel constituent	Composition (%)
H_2O	0
C	--
H_2	24
CO	40
CO_2	25
CH_4	5.6
C_2H_x	2.4
Other hydrocarbons	2
H_2S	0.05
N	--
CL	0
O	0
Ash	0
S	--

a. Source: Union Carbide Corporations.

Two residue streams are produced by the Purox[TM] system--a molten slag or quenched frit stream that exits from the bottom of the reactor, and a wastewater stream resulting from the gas cleaning process train. The frit consists of the ash, metals and glass contained in the original feed material. It passes through the combustion zone, where it is heated to a molten state and sterilized and then quenched in a water bath. It can be either landfilled or used as fill material. The ash contains only 2% of the volume of the in- coming refuse. The waste water stream resulting from the gas cleaning train contains soluble organics, which result in a chemical oxygen demand loading of 60,000 to 70,000 mg/ liter. These are pretreated on site before being released

TABLE VIII

Comparison of Fuel Gas, Methane, and Propane[a]

Parameter	Purox fuel gas	Methane	Propane
Lower heating value (BTU/SCF)[b]	350	910	2,312
Compression power (KWH/MM BTU)[c]	5.7	1.8	0.6
Combustion air requirements (SCF/MM BTU)[d]	8,300	10,500	10,300
Volume of combustion products (SCF/MM BTU)	10,500	11,600	11,200
Heat release/volume of combustion products (BTU/SCF)	95	86	89

a. Source: Union Carbide Corporation.

b. Standard cubic foot dry, as measured at 60° F and 1 atm pressure. Heating value calculated at 18° C.

c. Gas compressed to 35 psig from 1 atm, 100° F, with 75% efficiency.

d. Based on the air needed to convert the fuel to CO_2 and H_2O--no excess air.

to the municipal waste water system. Figure 2 presents the mass balance of the oxygen conversion system.

B. The Purox[TM] System and Blended Refuse

The problems inherent in the use of MSW include efficiency, erosion and pollution generated by the inorganics contained in the waste stream. Wood waste faces a significant problem of energy efficiency due to high moisture content. Blending in the feedstocks suggests that the mixture could minimize both sets of problems. The use of the Union Carbide system further addresses these problems.

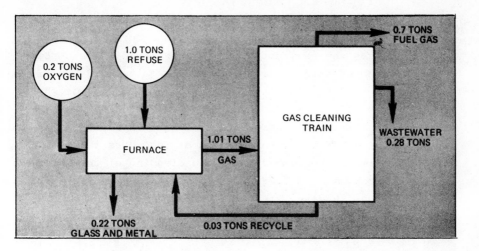

Fig. 2.   Inputs and products of Purox system.

1. Energy efficiency considerations

Since system efficiency is affected by the presence of both inorganic material and water, these problems can be treated together. Regarding the presence of metals and glass, Union Carbide has operated its Charleston, West Virginia, facility on both as-received refuse and on refuse that has been shredded and magnetically separated. In the latter case, the inorganic constituents entering the reactor constituted approximately 18% of the feed materials. Gas production increased slightly, while gas quality improved significantly. Union Carbide officials theorize that the increased ratio of organics to inorganics decreased the pyrolysis temperature slightly, increasing the production of C_2 to C_6 light hydrocarbons to 4.8% of the gas. This raised the heating value of the gas to 345-370 BTU/scf. Blended waste will provide the reactor with a feed consisting of 87% organic material and 13% inorganic matter. Such a ratio could increase the volume of gas produced, as well as the heat content of that gas.

Water presents the second energy efficiency considera-tion. By blending the waste streams, the as-received mater-ial could have an average moisture content of about 35%.

This is a useful feedstock. The drying zone reduces this
amount of water further before pyrolysis occurs, although
this requires the consumption of more oxygen. Union Carbide
Corporation estimates indicate that it requires 0.2 lbs of
oxygen to produce 1 lb of feed material, regardless of what
that feed material is. A ton of feedstock with 50% water
will necessitate consumption of 200 lbs of expensive oxygen
just to drive off the water. A ton of feed material with
35% moisture--as contemplated in the blending of untreated
MSW and wood waste--will require 140 lbs of oxygen for drying purposes. However, the wood waste can be dried to 30%
moisture before it is mixed with MSW and fed into the
reactor--at a cost of 250,000 BTU per ton of blended feedstock. The resulting reactor feed will contain about 25%
moisture. The oxygen consumption in handling this partially
predried feed material will be only 100 lbs/ton. Predrying,
then, can produce significant operating cost savings.

2. Erosion and pollution considerations

Erosion, from the point of the fuel user, is eliminated because the energy product delivered is a clean fuel
gas. The system operator, by utilizing belt- or pan-type
conveyors, does not face severe erosion of the materials
handling systems. And, since the hearth of the Purox^{TM}
reactor is both refractory lined and water cooled, a layer
of fused slag builds up in the hearth, preventing severe
erosion of the energy production system.

Potential pollution problems from this system appear
only in the waste water, since the gas is clean and the frit
is sterile. The fly ash, which presents problems in direct
combustion systems, ends up in the frit. The product gas
contains considerably less particulate than the EPA
standard.

IV. CONCLUSION

The mixing or blending of municipal and wood wastes
suggests both economic and technical advantages for the
production of fuel from waste. Such advantages include the

increased tonnage of material available and the improved
quality of the feed material. By using the Purox^{TM} process,
a highly desirable fuel can be produced that enjoys broad
market applications.

The final advantage of such blending stems from the
market consideration; the presence of wood wastes indicates
that strong markets are available for the fuel. Such operations as kiln drying of lumber, veneer drying, pulping, and
paper making require significant amounts of energy. Further,
they are 24 hr/day operations that may require gaseous fuels.
Thus, they are ideal for fuel-gas production systems. Thus,
the waste disposal system can be tied back into the wood
products industry, an industry with long traditions of using
novel and innovative fuels. In making such a link--in the
marketing of the fuel--the Northeast region wood products
industry could increase its energy independence from the
present level of 11%.

Blending of wastes is effective, but it is not a uni-
versal problem solution. For example, wood waste and animal
waste may be incompatible. , Both offer water problems;
manure may contain 90% moisture. Thus, the mixing of these
two wastes may create a feedstock that is too wet to convert
into gas by thermal processing. At the same time, the lignin present in the wood waste may inhibit biological conversion to methane. Blending urban and animal waste may preclude thermal but not biological conversion to energy.

These limitations do not rule out mixing of wastes.
The case for blending urban and wood waste offers counter-
vailing evidence. What those limitations illustrate is that
each particular mix must be examined in terms of the feedstock created and the range of alternative conversion
processes that will be most amenable to the handling of that
feed material.

ACKNOWLEDGMENT

The author wishes to acknowledge the assistance of
Dr. Bernard D. Blaustein, Pittsburgh Energy Research Center,
in the preparation of this manuscript.

REFERENCES

1. Kyosti V. Sarkanen, "Renewable Resources for the Production of Fuels and Chemicals," Science, Vol. 191,
   Feb. 20, 1976, p. 773.
2. National Materials Advisory Board, "Problems and Legis-
   lative Opportunities in the Basic Materials Industries,"
   National Academy of Sciences, 1974, p. 54.
3. D. A. Tillman, Particleboard Raw Material Availability
   Report, New Hampshire Department of Resources and
   Economic Development, Dec. 18, 1970, p. 2.
4. Edward P. Cliff, Timber: The Renewable Material,
   National Commission on Materials Policy, August, 1973,
   pp. 3-8.
5. M. D. Schlesinger, W. S. Sanner, and D. E. Wolfson,
   "Pyrolysis of Waste Materials From Urban and Rural
   Sources," Proc. Third Mineral Waste Utilization Symp.,
   March 14-16, 1972.
6. J. E. Anderson, "The Oxygen Refuse Converter--A System
   for Producing Fuel Gas, Oil, Molten Metal and Slag From
   Refuse," Union Carbide Corporation.

PYROLYSIS OF PINE SAWDUST*

J. A. Knight

Engineering Experiment Station
Georgia Institute of Technology
Atlanta, Georgia

I. INTRODUCTION

The utilization of waste materials is of vital concern in the United States because these materials--
agricultural, forestry, and municipal wastes--represent
unused resources and in many cases present serious disposal
problems. During the past several years, a great deal of
attention has been given to pyrolysis as a means of converting these materials into useful products, particularly fuels.
Workers at the Engineering Experiment Station (EES), Georgia
Institute of Technology, have been utilizing the pyrolysis
approach to convert agricultural and forestry wastes into
char, pyrolytic oil and combustible gases [1-3]. Particular
attention is being given to forestry wastes at the EES.

---

*Supported in part by EPA Grant No. R 804416010.

These materials are abundant and in many cases are generated
in large concentrated quantities at sites such as sawmills
and paper mills and represent a relatively homogeneous material. In the United States, it has been estimated that 100
million dry tons of forestry wastes are generated annually
[4]. Work with forestry wastes at the EES has been mainly
with pine bark and sawdust because of their importance to
the southeastern states. The EES has two continuous pyrolysis pilot plants that are used in research and development
work in the pyrolysis of all types of lignocellulosic materials and municipal refuse. One of the pilot plants will
process 1500 lb of dry feed per hour, and the other system
will process 800 lb of dry feed per hour. The experiments
reported in this paper on the pyrolysis of sawdust were conducted to provide data mainly on the yields of products, the
heating values of the products, and compositions of noncondensible gases over the temperature range of 540°C to 870°C.
These data are useful as guidelines in the operation of the
two continuous pilot plant systems.

II. EXPERIMENTS

A. Feed Material

     The feed material was pine sawdust and was dried to
6% moisture on a wet basis. The sawdust on a dry basis had
1.2% total ash. 0.3 acid insoluble ash, and a heating value
of 8103 BTU/lb. The bulk density of the sawdust was 14.1
lb/ft^3, and 20% was +4 mesh, 68.0% was 4 x 16, 10.4% was
16 x 35, and the remainder (1.6%)was less than 35 mesh.

B. Pyrolysis

     The pyrolysis experiments were conducted in a Lindberg
Hevi-duty, 6 inch tube furnace (Fig. 1) fitted with a stainless steel tube (type 306). The furnace has three sections,
which can be heated independently. Thermocouples were
placed in the charge and the exit gas space. A charge of
2500 gm sawdust, 6% moisture on a wet basis, was held in
place in the center of tube with spacers. The furnace was
heated until the evolution gases had essentially ceased. The

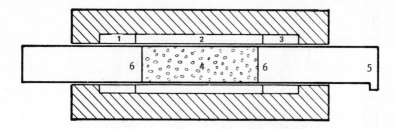

Fig. 1.   Lindberg Hevi-Duty furnace; (1-3) furnace
thermocouples, (4) thermocouple in center of charge,
(5) thermocouple at gaseous exit.

temperatures of the furnace, in the center of the charge and
at the gaseous exit, are given in Table I.

The temperature at the exit end of the tube was high
enough so that condensation did not occur. The gas evolution
rate for each experiment showed a rapid increase when the
center of the charge reached the 150-175°C range and continued to evolve at a fairly rapid rate until the maximum temperature was reached. The off-gases were passed through a
condensation train (Fig. 2) so as to condense the organic
and aqueous material, and the gases were collected in large
plastic gas-collection bags. Total volume of evolved gases
were measured with a dry test meter.

At the end of a pyrolysis experiment, the ball valve
(Fig. 2) was closed, and the char, after cooling to room
temperature, was removed, weighed, and samples prepared for
analysis.

TABLE I.

Maximum Temperatures During Pyrolysis

Location	Temperature (°C)			
Furnace	540	650	760	870
Middle of Charge	500	610	700-740	720-810
Gaseous Exit	230	280	320	360

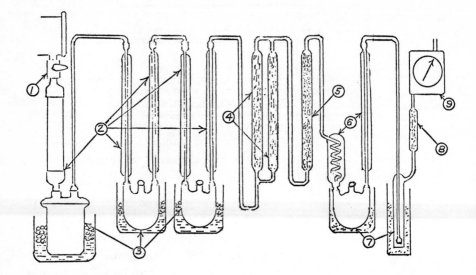

Fig. 2.   Condensation train; (1) stainless steel ball
valve, (2) water-cooled ball condensers, (3) ice-cooled
traps, (4) glass wool demisters, (5) Drierite column, (6)
cold condensers, (7) dry ice traps, (8) magnesium perchlorate dryer, and (9) dry test meter.

C.   Condensation Train and Collection of Organic and Aqueous
     Phases

     The off-gases from the pyrolysis tube pass through a
stainless steel ball valve   (1) [in Fig. 2] and then through a
series of water-cooled condensers (2) and ice-cooled traps
(3) (See Fig. 2). A jacketed stainless steel tube is used
for the first condenser in place of a glass condenser. The
use of a resin kettle as the first trap facilitates the
recovery of viscous, oily condensates. The gas stream then
passes through the two parallel glass-wool demisters (4) and
a calcium sulfate (Drierite) column (5) into a series of cold
condensers (6) and cold traps (7). The purpose of the demisters is to remove oil mist, and the calcium sulfate is used
to remove the last traces of moisture from the stream. The
cold condensers (6) are chilled with ethanol, which circu-
lates through a heat-exchanger coil immersed in dry ice and

acetone, and the traps (7) are immersed in a bath of dry ice and acetone. From the cold traps, the gases pass through a magnesium perchlorate dryer (8) and a calibrated dry test meter (9) into a series of 96 liter gas-collection bags. The use of magnesium perchlorate has been found to present fouling of gas chromatographic columns during analysis of the noncondensible gases. The quantity of magnesium perchlorate is held to a minimum, and precautions are taken to prevent any explosion hazard. At the completion of an experiment, the weights of the condensates were stored in tightly closed containers. Heavy organic and aqueous condensates were stored at ambient temperature, and the condensates from the dry ice traps were stored in a freezer.

D. Noncondensed Gases

During the pyrolysis experiment, the gases were collected in 90 liter quantities successively in 96 liter gas-collection bags. After each bag is filled, it was kneaded to mix its contents, and approximately half of its contents were aspirated through gas-collection tubes (approximately 250 ml capacity). Two gas-collection tube samples were taken for analysis.

E. Analytical Procedures

1. Heating values

The heats of combustion of the feed material, chars, and organic liquids were determined in a Parr bomb calorimeter using the ASTMD-240-64 method.

2. Water in pyrolytic oils and aqueous phases

The percentage of water in the pyrolytic oil phases and aqueous phases was determined by a 2 hr distillation with toluene in a Dean-Stark apparatus. The percentage of water in the pyrolytic oils was generally less than 25%, and the percentage of water in the aqueous phase was generally greater than 75%. The ASTMD-95-70 (modified) was followed in these determinations.

3. Volatiles in char (Low temperature)

The weight loss in the char was determined by drying a weighed sample for 2 hr at 105°C followed by cooling 1 hr in a vacuum desiccator.

4. Volatiles in char (High temperature)

The volatiles in a char sample were determined by heating the sample from the 105°C determination at successive temperatures as follows: 2 min at 300°C, 3 min at 500°C, and 6 min at 900°C. The sample was cooled in a vacuum desiccator and then weighed to determine weight loss, which was calculated as percentage of volatiles.

5. Total ash

The char sample from the volatiles determination was ignited without the crucible cover to constant weight at 550-600°C, cooled to room temperature, and weighed.

6. Gas analysis

The gases were analyzed utilizing four gas chromatographic units, and the conditions are given in Table II.

III. RESULTS AND DISCUSSION

The yields of the four major pyrolytic products versus temperature are presented in Figure 3. The char yield decreases with increasing temperature and levels out at about 760°C. The yield of the combined oil condensates does not show any noticeable change with temperatures, except for the value at 760°C. This higher value may be due, in part, to an experimental error. The yield of the combined noncondensible gases increases with temperature, and the water yield decreases with temperature. The sums of the yields of all the products are within experimental error and show that essentially all of the products were recovered. These data are useful as guidelines in the operation of the two continuous pyrolytic pilot plants at the Engineering Experiment Station. One pilot plant has a capacity of approximately 1500 dry pounds per hour, and the other 800 dry pounds per hour.

TABLE II

Gas Chromatographic Analytical Conditions

Gas component	G.C. column	Temp.(°C)	Flow rate; carrier gas	Detector and Instrument
Hydrogen	Silica gel; 60 x 80 mesh, 6' x 1/8"	32	18 ml/min; argon	Thermal conducti-vity; Carle AGC
Oxygen, nitro-gen, methane, carbon monoxide	Molecular sieve, 13 x; 60 x 80 mesh, 12' x 1/8"	40	20 ml/min; helium	Thermal conducti-vity; Perkin Elmer 990
Carbon dioxide	Activated carbon; 60 x 80 mesh, 10' x 1/8"	150	22 ml/min; helium	Thermal conducti-vity; Carle AGC
Hydrocarbons (C_2 to C_5)	Alumina F-1; 60/80 mesh	50-150 at 12°/min.	24 ml/min; nitrogen	FID; Perkin Elmer 800

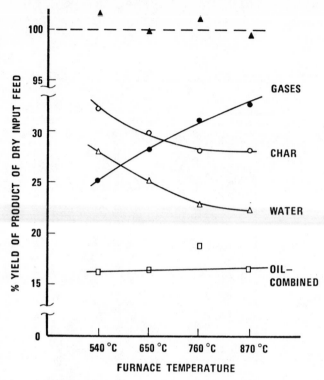

Fig. 3. Yields of the four major pyrolytic products versus temperature.

Analysis of the noncondensible gases showed that major gases are hydrogen, methane, carbon monoxide, and carbon dioxide. The C_2, C_3, and C_4 hydrocarbons are produced in relatively small amounts, and no experimental attempt was made to identify the individual components of these hydrocarbons. The variations of the yields of these gases are shown in Figure 4. The carbon monoxide and carbon dioxide both show significant decreases in yield with increasing pyrolysis temperature. The hydrogen yield shows a dramatic increase for the temperature range investigated. The methane yield increases to a maximum at approximately 680°C and then shows a gradual decrease at the higher temperatures. These data serve as guidelines for the expected yields of the major gaseous components in the operation of the EES pilot plants. Based on the gas composition, heating values of the gaseous

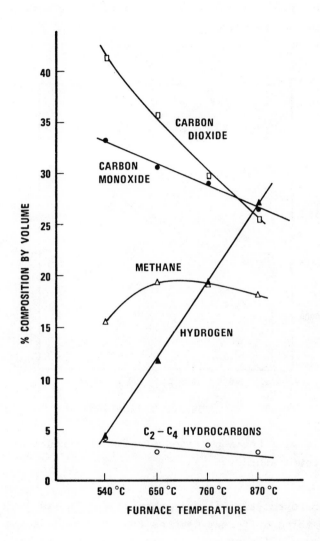

Fig. 4.   Variations of yields of the major gases.

mixture produced at each temperature were calculated, and
these are presented in Figure 5. The heating values of the
pyrolytic gas increases with increasing temperature and shows
a range of 350 to 410 BTU/ft^3. In actual practice in a continuous pyrolysis system, the heating value of the gas will
be decreased by nitrogen from process air.

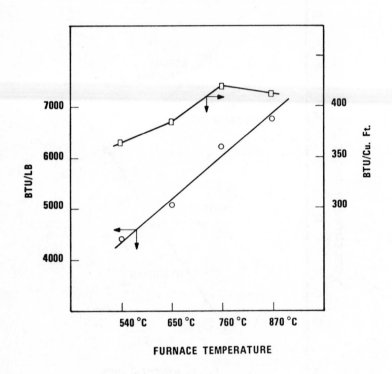

Fig. 5. Heating values of gaseous mixture produced at
various temperatures.

The pyrolytic oil from the off-gas stream was collected into three different samples from the condensation
train. The yields of each of these oils are presented in
Figure 6, along with the combined yield. The reason for the
higher yield values obtained for condensates 1 and 2 is not
known. It is felt that they may be attributed in part to
experimental error. Oil condensate 1 obtained from the first
condenser represents the largest fraction and shows that the

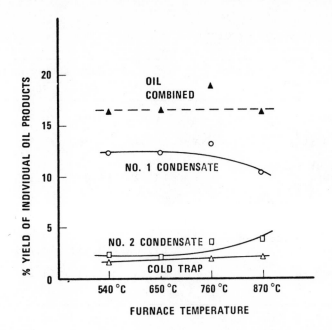

Fig. 6.   Yields of pyrolytic oil from the off-gas
stream.

oil yield decreases as the temperature decreases. Oil condensate 2 obtained from the next two condensers shows a
slight increase with increasing temperature. The oil yield
represented by the cold trap was condensed from the stream in
a dry ice trap and was a relative volatile material. In
actual practice in a continuous pilot plant operation, this
oil would not be condensed from the off-gas stream and would
add to the heating value of the gas. The heating values of
the oils are given in Table III and are in the 12,000 to
14,000 BTU/lb range on a dry basis. In the operation of our
continuous pilot plant pyrolysis units, the general practice
is to operate the condenser so that the pyrolytic oil does
not have more than 15% water. The remaining noncondensed
gases, including water vapor, are burned in the afterburner.
        The heating values of the pyrolytic products obtained
at each temperature are given in Table III. In Figure 7, the
heat available from each product versus temperature is

TABLE III

Heating Values of Pyrolytic Products

Product	Furnace temperature			
	540°C	650°C	700°C	870°C
Gas (BTU/lb)	4,411	5,029	6,205	6,771
Gas (BTU/ft^3)	348	362	395	409
Oil, first condensate (BTU/lb)	13,529	14,072	13,153	14,004
Oil, second condensate (BTU/lb)	12,158	13,555	12,637	12,716
Oil, dry ice condensate (BTU/lb)	13,362	13,643	13,290	13,432
Char (BTU/lb)	14,357	14,441	14,218	14,053

presented. The significant information to be gained from
this graph is that the sum of the heats available from each
product at each pyrolysis temperature is very close to the
heat content of the dry input feed material. These data show
that the total heat content of the input material is available for use in the three fuels--char, oil, and gas-- that
are produced in the pyrolysis process. In Figure 8, the percentage recovery of each of the pyrolytic products of the
input feed is presented. These data show that the heat
values in the char and oil decrease with increasing temperature with more gas being produced. At 540°C, for example,
the char and first oil condensate present 77.5% of the total
heat value of the input feed, whereas at 870°C, they represent 66.7%. In order to obtain the maximum recovery of the
heat content of the products, the gases must be utilized on
site. A portion of the gases can be used for drying input
feed, as the process operates best using feed material of
less than 10% moisture.

Additional analytical data on the char samples is
presented in Table IV. The significant information from this

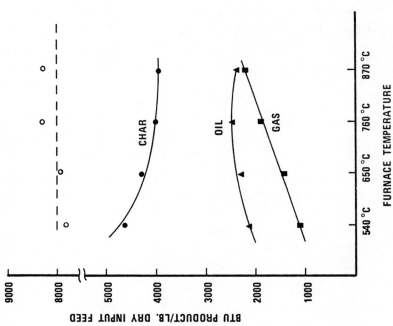

Fig. 8. Percentage of recovery of the pyrolytic products of the input feed.

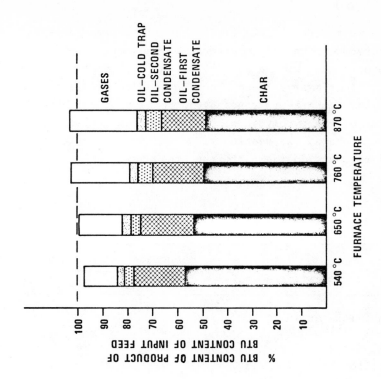

Fig. 7. Heat available from pyrolytic products.

table is the total volatiles in the char. The data show the temperatures that the char must be subject to in order to produce a char with a particular volatile content.

In summary, these bench scale pyrolysis experiments over the temperature range 540°C to 870°C provide data on the yields of the pyrolytic products, the composition of the gaseous phase, the heating values of the products, and the heat recovery in the different products.

TABLE IV

Char Analytical Data

	Furnace temperature			
	540°C	650°C	760°C	870°C
Weight loss at 105°C (2 hr) (%)[a]	1.1	0.5	0.6	1.0
Total ash (%)	2.3	4.6	3.6	3.9
Volatiles (%)	13.7	7.1	3.7	2.8

a. Weight loss probably due to vapors absorbed during cooling of char after pyrolysis.

REFERENCES

1. J. A. Knight et al., "Pyrolytic Conversion of Agricult-
    ural Wastes to Fuels." Presented at 1974 Annual Meeting
    of American Society of Agricultural Engineers, Oklahoma
    State University, Stillwater, Oklahoma, June 23-26,
    1974.

2. J. A. Knight and M. D. Bowen, "Pyrolysis--A Method for
    Conversion of Forestry Wastes to Useful Fuels,"
    Presented at Southeastern Technical Division of Ameri-
    can Pulpwood Association, Atlanta, Georgia, Novem-
    ber 5-6, 1975.

3. J. A. Knight and M. D. Bowen, "Pyrolysis--A Method for
    Conversion of Forestry Wastes to Useful Fuels."
    Presented at Workshop on Wood Residues: An Alternate
    Energy Source for New England Industry, University of
    New Hampshire, Durham, New Hampshire, April 7-8, 1976.

4. Heyward Taylor, Division of Engineering, Forest Service,
    U. S. Dept. of Agriculture. Private communication.

THE ENERGY PLANTATION: DESIGN, OPERATION

AND ECONOMIC POTENTIAL

J. F. Henry, M. D. Frazer, and C. W. Vail

InterTechnology Corporation

Warrenton, Virginia

I. INTRODUCTION

The idea of using biomass, in the form of agricultural
and forest crops and their residues, as fuels or raw materials for chemical processes is not new. Indeed, at present,
wood products still constitute a major source of fuel in
many countries. In recent years, the crisis resulting from
the scarcity and increased cost of fossil and nuclear fuels
has motivated innovative research in the area of alternative
sources of energy. Biomass in its various forms is an
attractive source of energy or feed stock for industry, as it
collects and stores low-intensity solar energy through photosynthesis. Plant matter is a versatile matter that can be
used directly as solid fuel, transformed into a variety of
synthetic fuels or used as raw material for industry. It is
a renewable resource that can be produced and used with

little or no detrimental environmental impact. The Energy
Plantation^{TM} concept developed by InterTechnology Corporation
is a systematic and innovative approach to the production of
economic fuel from plant matter on a large scale. The following discussion will deal mostly with the production of biomass fuels, but essentially all the results can be applied to
the production of raw material for industrial applications.
The Energy Plantation^{TM} concept is a practical approach to
producing fuel from plant matter in that by choosing the
planting density and harvesting schedule of the most appropriate plant species for a plantation site, the capture and
conversion of solar energy is optimized.

The plantation is designed and operated to minimize
the cost of the plant matter produced. At present, Energy
Plantation fuels can be produced at costs competitive with
those of conventional fuels in many parts of the country.
Plant matter fuels have numerous advantages. Like fossil
fuels, they are a natural storage system from which energy
can be recovered at any time. The energy can be released at
high temperature and, thus, has high thermodynamic availability. Biomass fuels contain very little sulfur, which
helps eliminate pollution problems. The combustion of these
fuels does not modify the carbon dioxide or thermal balance
of the earth as burning fossil fuels does. None of the
potential hazards, high capital cost, and waste disposal
problems associated with nuclear fuels exist for Energy Plantation fuels. Finally, plant matter fuels are essentially
inexhaustible. The following discussion will consider the
biomass resources, land resources, biomass production rates,
design, operation, and economics of the plantation.

II. BIOMASS RESOURCES

A. Managed Plantations

An obvious way of generating plant matter for fuel or
raw material is to harvest wild vegetation growing on land
unsuitable or unused for agriculture or other purposes. No
planting or managing costs would be incurred, and the land
would be accorded some usefulness. However, depending on

wild vegetation has a number of disadvantages. First, annual
yields of plant matter per acre from natural stands are often
only half, or even as little as a quarter, of what can be
achieved in managed plantings [1]. Higher yields are possible with management, because planting density or spacing
between plants can be chosen to maximize the product of the
yield per plant multiplied by the number of surviving plants
for a particular harvest schedule. Moreoever, natural
growth is usually a mixture of many species, which produce
plant material at different rates; in a managed planting,
high-yielding species can be chosen that are well adapted to
the site. In a wild stand, even where particular species
predominate, the ages of individual plants usually vary considerably. The resulting variation in plant size makes
mechanized harvesting operations difficult, and hence costly.
Under management, the age of the plants, and hence also their
size, form, and structure, can be kept uniform, thereby
facilitating mechanical harvesting and processing.

B.   Annual Plants versus Perennials

Qualitative analysis indicates that perennial species
are to be preferred over annuals for several reasons. First,
growing annuals creates troublesome manpower and equipment
utilization problems. Annuals must be started at about the
same time every year when soil moisture and other conditions
are optimum for the species being planted. They must also
be harvested during a relatively short period after they
mature. Thus, much of the field machinery and manpower
needed are in use only a few days a year, unless arrangements
are made to move the machinery and operators from site to
site as the planting and harvesting activities move up and
down the country.

Although yields from annuals can be high enough to be
attractive, high yields are achieved only on land that can be
given intense soil preparation before planting, which means
land that is in demand or may come into demand for food production. Growing fuel on this land is not likely to win
ready public support. Another disadvantage of annuals is the
high cost of producing plant matter from them. Biomass from

annual crops, being harvested all at once at the same time,
would have to be stored between harvests and preserved
against biological degradation during storage. The preservation and proper storage provisions would undoubtedly be
costly.

In contrast to annuals, perennials can be harvested
more or less continuously through the year in response to the
demand for fuel. The harvesting machinery, transportation
equipment, and personnel involved can therefore be programmed
for more continuous use. Consequently, a permanent equipment
and work force can be established for these cyclical operations in the plantation. The preservation of perennial crops
is accomplished by nature until the plants are harvested as
long as they are alive in the field.

C.   Deciduous Plants versus Conifers

Qualitative analysis as well as quantitative data on
plant growth rates indicates that deciduous woody species are
preferable to conifers for Energy Plantation^{TM} culture. Data
on plant-matter production rates indicate that conifers generally grow more slowly in shorter rotations than deciduous
species. An additional advantage of deciduous species is
that many will sprout vigorously from their stumps after
their structure above ground has been harvested. Since
planting costs are a substantial part of the costs of producing any crop, those deciduous species that sprout readily
from their stumps and hence provide several crops per planting have an advantage over other species that do not. Few
conifers sprout after they have been cut down.

Some deciduous species which grow rapidly when they
are young and sprout vigorously from the stumps after being
harvested can also be reproduced vegetatively from clones. A
clone is a live stick cut from a plant, which, after storage
under moist and cool conditions, will start growing rapidly
soon after planting. This is another trait not shared by
conifers. Vegetative reproduction has numerous advantages
over reproduction from seeds. First, it is far easier and
cheaper to collect clones from trees than seeds. Second,
clones produce a plant genetically identical with the one

from which they were cut, whereas seeds may not. Clones can
be collected from plants already growing in the plantation,
making a nursery unnecessary, the clones being already
adapted to local plantation conditions. They can be planted
directly in the land where the plants are to be grown, making
transplanting unnecessary. It is evident from this list of
advantages of using clones that the costs of collecting and
planting clones will be far lower than the costs of collecting and germinating seeds and then transplanting the seed-
lings to the plantation site.

D. Preferred Deciduous Species

A number of well-known deciduous species have these
desirable characteristics. Some varieties are members of
hybrid "families" developed for propagation in a wide variety
of soil types and climates. The hybrid poplars are an
example of such a "family." Other desirable species are natural ones having several varieties that with knowledgeable
selection, are suitable for a diversity of soil and climate
conditions. The cottonwoods are an example.

A list of deciduous species having the desirable
traits for Energy Plantations[TM] and where they have been
shown to grow satisfactorily at high planting densities on
short and repeated harvest cycles is shown in Table I. At
least some yield data are available for all the species at the
sites indicated by an X in the table. The important conclusion to draw from the table is that there is at least one
deciduous species that has been shown to grow well under
plantation-type conditions essentially everywhere in the
United States where establishing plantations for fuel production may be of interest. This list of species probably
suitable for plantation culture presented in Table I is
probably not exhaustive. On the contrary, it is likely to be
biased by the purposes for which the species listed originally became of interest. In no case has that interest been
producing fuels. It is possible that species exist that are
more satisfactory for fuel production but for which little
or no yield data have been recorded.

TABLE I.

A Representative List of Deciduous Species That Show Promise for Plantation Culture

State	Hybrid poplars		Other species											
	NE 388, 49 and 252	Others	Aspen and hybrids	Black cottonwood	Red alder	Sycamore	Pin cherry	Plains cottonwood	Eastern cottonwood	Silver maple	European black alder	Green ash	Sweetgum	Eucalyptus
New Hampshire			X				X							
Wisconsin		X	X											
Minnesota			X											
North Dakota								X						
Washington			X	X	X									
Pennsylvania	X	X												
Ohio											X			
Indiana									X					
Illinois									X		X			
Nebraska									X					
Kansas								X	X	X		X		
Georgia						X						X	X	X
Alabama									X			X	X	
Mississippi						X								
Louisiana									X					
Texas									X					
Florida														X

E.   Preferred Perennial Grasses

Certain perennial grasses are also promising sources of raw material for fuel production. Most of these grasses can be reproduced vegetatively, and they regrow rapidly after harvest. They are similar in these respects to the deciduous tree species previously discussed. More than one harvest can usually be reaped from them every year, but the actual number depends on the length of the growing season, the regularity and amount of rainfall, and the ambient temperatures during the growing season. There are two broad categories of perennial grasses, the cool-season grasses and the warm-season grasses.

In those parts of the country where frosts occur every winter, and hence, where only cool-season grasses will grow,

annual yields are 3 to 5 tons of oven-dry biomass per acre
[2]. Such yields are too low to be practical for large-scale
fuel production, and moreover, cool-season grasses present
many of the rather serious problems previously associated
with annual crops. Nor are cool-season grasses good candidates for those parts of the country that are usually frost-
free the year round.

In contrast to cool-season grasses, the growth rate of
warm-season grasses does not decline in the warmest months.
In fact, providing there is sufficient soaking rain (2 or 3
inches per month), their growth rate increases as the temperature rises to its peak in the summer. In Florida and
localities around the Gulf, yields between 8 and 10 tons per
acre per year of oven-dry material are attainable in managed grasslands [3, 4]. Certain warm-season grasses are
therefore promising candidates for Energy Plantations^{TM} in
parts of the south. Promising warm-season grasses include
perennial sorghums, sugarcane relatives, bamboo relatives,
bahia grass, and finally, the Bermuda grasses, which appear
to be the most attractive candidates.

III. LAND RESOURCES

A.   Land Suitable for Energy Plantations^{TM}

First of all, to be suitable for growing Energy
Plantation^{TM} crops in high yield, land must have an adequate
supply of water. The amount of water required varies among
species from somewhat less than 200 to somewhat more than
300 pounds of water per pound of oven-dry plant matter produced [5]. To be of practical interest, a combination of
site quality and crop management must produce at least 7
tons, and preferably nearer 10 tons, of harvestable, oven-dry
plant material per acre per year. The combined effect of the
water and harvest-yield requirements means that practical
plantations cannot be established in territories where precipitation is normally less than about 20 inches per year.

In the contiguous forty-eight states, precipitation is
generally at least 20 inches per year in the territory east
of about the 101st meridian and on the western slopes of

the mountains along the Pacific coast. The land between these
two regions, except for the western part of Idaho and eastern
Washington, normally experiences less than 20 inches of precipitation per year and is therefore generally too arid for
worthwhile plantation operation.

The steepness of slopes in the terrain is a second
factor that influences the practicality of Energy Plantations[TM]. Hilliness rules out most of the land on the western slopes of the mountains along the Pacific coast, nearly
all the land with more than 20 inches annual precipitation in
Idaho and eastern Washington, and the land in the Appalachian
region in the east.

The third factor in determining land suitability is
population density. Counties having population densities in
excess of three hundred persons per square mile have been
excluded from consideration for Energy Plantation[TM].

In general, land meeting these criteria is within the
eastern and central time zones, excluding the land in the
hilly Appalachian region and densely populated areas.

B.   Availability of Land

Whether land that is suitable for Energy Plantation[TM]
according to the criteria previously described can be made
available for this purpose depends upon both the current use
of the land and its capability for sustaining plant growth.
Land that is currently used for a valuable purpose such as
cropland, commercial forest, pasture, range, or recreation is
unlikely to be made available for plantations. On the other
hand, there are large areas of land that cannot be used for
these purposes or that perhaps are only marginally useful,
but that can be used for Energy Plantations[TM]. For example,
land that is unsuitable for plowing and other intense soil
preparation procedures required for agriculture can be used
for plantations of woody perennial crops. Land with slopes
as steep as 25% appears to be acceptable for plantations
where planting is to be done with clones [6].

The natural fertility of the soil at potential plantation sites is less important than the ease with which its
fertility can be maintained at a high level of fertilization.

The major source of fertilizer will be ash or other residues
from the fuel-using or conversion operation. Excessively
porous soil will be unattractive. On the other hand, poorly
drained soils will be undesirable because of their adverse
effect on plant growth and the difficulties they create with
field machinery. Sites with a sufficiently loamy soil, which
will support field machinery and retain adequate moisture,
are preferred for deciduous plantations. The depth of the
surface soil layer is expected to be less critical for deciduous plants than for annual farm crops, because the roots of
the former will penetrate deeper and through more resistant
soil structures than those of the latter.

Based on considerations of land capability and land
use--both potential and current--an estimate has been made of
the suitable land in the contiguous forty-eight states that
might become available for Energy Plantations^{TM}. This land
area is estimated to be of the order of 100 million acres.
This estimate includes only relatively large tracts of land.
Many unused small plots are available, which, if the economics of plantation operation on a cooperative basis are favorable, could boost up the estimates of the land area available for Energy Plantations^{TM}. Our estimate of land available for Energy Plantations^{TM} was recently confirmed by the
Department of Agriculture, which evaluated the available land
area at 110 million acres [7]. All of this land is believed
to meet the site, climate, and precipitation requirements for
deciduous species preferred for plantation culture. However,
only that part of the estimate representing land in the
vicinity of the Gulf of Mexico will be suitable for warm-
season grasses; it is estimated that 25 to 30 million acres
meet their requirements.

It should perhaps be emphasized that this 100 million
acres has no particularly valuable potential or current use;
it is not prime cropland, commercial forest, pasture, range,
or recreational land. Virtually all of it is privately owned,
and its owners would probably welcome the opportunity to
receive a steady income if such land were to be devoted to
Energy Plantations^{TM}.

IV. BIOMASS YIELDS

A. Available Data

   To evaluate the amount of plant matter that could be
produced from woody perennials on a given plantation site,
estimates of the rate at which biomass can be grown on the
area are needed. The plant-matter production rate per acre
varies with the number of surviving plants per acre and the
age of the stand at harvest time. Other factors affecting
the sustained annual yield per acre include the species
planted, the cultivation and fertilization programs, the
local climate and soil character, and the number of harvests
that can be obtained from a plantation before it has to be
replaced.

   Means are needed for predicting the expected annual
yields of plant matter at a particular site. The procedures
used in the forest products industry for estimating pulpwood
and lumber yields are not suitable for estimating yields from
an Energy Plantation^{TM}. These methods are not designed for
estimating growth rates year by year in the first 3 or 4
years of growth, nor are they designed for estimating regrowth
from stumps. Therefore, a yield-estimating procedure has had
to be devised for deciduous species grown on short harvest
cycles, high planting densities, and repeated harvests. Sets
of data useful for devising a system for predicting plant-
material production rates are available for about fifteen
species and varieties grown in about as many sites in the
midwest and south. In a few instances, data are available
for a particular species at more than one site, and in
others, for several species at a particular site.

B. Growth Prediction Model

   It has been found that an effective way to correlate
the yield data is to express yields as the product of the
number of living plants surviving to harvest time and the
harvestable weight per plant at that time. Analysis of the
available survival data shows that the most important factors
influencing the number of surviving plants, other than the
particular species involved, are the age of the stand when it

is first harvested and the original planting density
expressed as the number of plants per acre. The survival
rate is also affected by cultural treatments and, to a lesser extent, by fertilization.

By regression analysis of the data, the numerical
constants needed for expressing the relationship between survival and the influential variables have been determined for
several species and varieties of interest for Energy Planta-
tion^{TM}, such as cottonwoods, silver maple, sycamore, and a
number of hybrid poplars. These relationships can be used
for predicting the survival of these species in plantations.
Further analysis allowed development of generalized means for
estimating survival when data are not available for specific
species-site combinations.

Analysis of the data available on the harvestable
yield per plant shows that the yields can be expressed in
terms of two constants, a growth parameter and a limiting
factor. Correlations have been established between these
constants and the planting area per plant that are valid in
the range of planting densities of interest. The influence
of several factors, such as species, cultivation, and fertilization, on the constants has also been elucidated for a
number of cases.

Being able to represent widely different growth data
in a systematic way by a small number of characteristic constants, as well as being able to generate correlations
between the characteristic constants and parameters of importance, such as the planting density, provides an important
tool for use in plantation design.

C. Application of the Growth Prediction Model

The system includes twelve relationships in addition
to identifying the species. A program for manipulating it by
computer has been written, and the system has been validated
by demonstrating its use for predicting yields from several
species at a variety of sites. Comparison of predicted
yields with actual yields leads to the conclusion that yield
predictions are good to probably $\pm20\%$ for a particular

species at a given site, but that for a group of species
adapted to a particular site, the yield prediction is
probably reliable to within about ±10%.

The average annual sustained yield from a plantation
is the sum of the yields from each of the harvests taken from
a planting divided by the number of years that elapse between
planting time and the first harvest and between harvests
after the first. The average annual sustained yield from a
species at a given plantation site is thus a complex function
of these three variables and the number of harvests taken
between plantings. Values of these variables that maximize
the average annual sustained yield from a species-site
combination can be determined from the model.

An example of the kind of analysis that can be performed with the growth prediction model is shown in Figure 1.
The estimated yield curves shown in this figure were calculated for a particular hybrid poplar at a site in Pennsylvania. Estimated yields are shown in the figure at two
planting densities and three ages at first harvest. A total
of six harvests is assumed to be taken from the stand before
it is destroyed and replaced by a fresh planting.

The yield curves of Figure 1 correspond to the harvesting of stem and branch wood material only. During part
of the year, leaves will also be harvested, and this will
result in a 5-10% increase in average annual biomass yield.
The annual yields indicated in the figure are significantly
higher than those of natural stands or forest products plantations and could cast some doubt as to the validity of the
growth prediction model. In fact, the yields reported in
the figure match very closely actual data recorded in
Pennsylvania [8]. Yields higher than those estimated in the
figure (10 dry tons per acre-year and over) have also been
reported in Canada [9] and Sweden [10] in experimental plots
with deciduous species of the genera Populus (poplar) and
Salix (willow). Yields in the range of 8-10 dry tons per
acre-year thus appear achievable on a sustained, large-scale
basis.

A number of species-site combinations have been
examined with the model, and in general, the highest annual

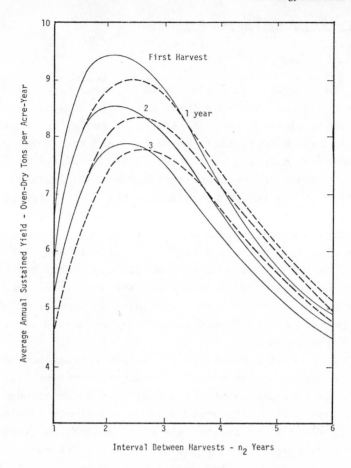

Fig. 1. Estimated average annual sustained yields from hybrid poplar NE-388 grown at Musser Farm, Central Pennsylvania [8]. (————) 4 ft^2 per plant, (- - -) 8 ft^2 per plant.

sustained yield per acre, which varies with the combination, is predicted at a planting density of 4-12 ft^2 per clone (10,900 to 3,600 clones per acre), with a harvest schedule in which the first harvest is taken when the stand is 1 or preferably 2 years old and subsequent harvests are taken at intervals of 2-4 years.

V. PLANTATION DESIGN AND OPERATION

A.   Plantation Operations

In an established plantation, operations are conveniently organized into four functions--supervision, field operations, clone production, and motor pool. Field operations include weed control; harvesting and transporting chipped plant material to the fuel-using facility; backhauling the ash and/or other residues to the plantation and spreading it (to return nutrients to the land); applying lime, fungicide, and eventually chemical fertilizer; and replanting. Clone production includes cutting stock from living plants, cutting the stock into clones, and packing and storing of clones. The motor pool is responsible for maintaining equipment, delivering fuels and supplies to the field operations, and controlling the reserve equipment. Plantation operations are assumed to be underway around the clock, seven days a week.

In a deciduous species plantation harvesting, transportation of the plant material, bringing residues back to the plantation, and maintaining equipment go on throughout the year at daily rates assumed to be relatively constant. The only seasonal operations are clone production, replanting, and weed control. Most of the seasonal work is done by part-time, relatively unskilled people. All other work at the plantation is done by full-time personnel.

In an established plantation where a warm-season grass is grown, the operations are conveniently divided into supervision, field operation, and motor pool. The responsibilities of these operations are similar to those of the corresponding operations in a deciduous plantation. An operation equivalent to clone production in deciduous plantations is not required.

Since warm-season grasses are practical fuel sources only in those parts of the country where the growing season for the grasses is between 9 and 10 months a year, grass will be harvested at least 9 months a year. Moving harvested grass to the fuel-using facility will be a year-round operation. Transporting and spreading residues on the plantation will also be year-round. Renewing grass stands, which will involve deep disking and possibly also plowing, will be a seasonal operation, a substantial part of which can be done between growing seasons. Root stock left in the ground will resprout to establish the new stand with some replanting probably required.

In warm-season grass plantations, large-capacity hay-making machinery will be used. The green hay will be windrowed as it is cut and left on the ground until it is approximately air dry. It will then be assembled into cylindrical bales, which will be stored in the open on the plantation for later shipment to the fuel-using facility.

B. Equipment Requirements

Most of the types of field and transportation equipment required for plantation operation will be standard equipment already available. The harvesting equipment for deciduous plantations, the harvester-windrower for warm-season grass plantations, and over-the-road trailers for moving plant material to the fuel-using facility and for bringing back residues to the plantations are not.

For moving plant material from the point where it is harvested or is in storage, over-the-road tractors pulling two semitrailers are assumed. The semitrailers will be equipped with live bottoms of the type used for manure spreaders, so they can be used for moving plant material to the fuel-using facility, for back-hauling residues, and after returning to the plantation also for spreading the residues on the land. The gross load capacity of the trailers will be about 16 tons.

The harvesters contemplated for deciduous plantations are not standard equipment. It is visualized that they will be similar to four-row silage corn harvesters, except that they will be built for rows 4 ft apart, and their guide chutes and plant-severing devices will be designed for handling up to four or five stems per chute at a time. The stems are expected to be between 1 and 2 inches in diameter and, on occasion, perhaps as much as 3 inches in diameter. The plant material to be severed, however, will always be

relatively juvenile and may not offer much more resistance
per stem than mature corn or sorghum. Standard two-row
silage corn harvesters have in fact been used on 2-year-old
coppice sycamore with good results. The capacity of the harvesters is assumed to be 8 tons (oven-dry basis) per hour.
The severed plant material will be conveyed to chippers
mounted on the harvester, and the chips will be propelled by
the chipper rotors to an over-the-road trailer traveling
alongside the harvester.

The capacity of the harvester-windrowers contemplated
for warm-season grass plantation is assumed to be about 7
tons (oven-dry basis) per hour, or about twice the capacity
of similar equipment regularly in use these days. Except for
the difference in capacity, the harvester-windrowers are
expected to be of essentially standard design.

C. Plantation Area Requirements

The estimated capital costs of harvesters and over-
the-road tractor-trailer units are substantial. It is therefore appropriate to select the area of a plantation production unit so that the potential work capacities of the most
costly units of equipment are fully utilized. In making
this selection, it is assumed that the harvesters and over-
the road transportation equipment are in use 90% of the time
they are in service. When they are out of service for major
maintenance, they are replaced by units carried in a maintenance reserve pool.

For a plantation where the maximum annual productivity
of plant material is 9 oven-dry tons per acre, the minimum
area that makes effective use of the harvesting and transport
equipment is estimated to be about 28,500 acres. It is
recognized that often it may be impossible to assemble an
area as large as this in a single tract. However, in localities having substantial areas potentially available for
plantations, it should be possible to assemble the minimum
area in a few relatively large blocks that are close enough
to allow them to be operated conveniently as a single production unit. There is no reason why the area of a production unit cannot be larger than the minimum, and a plantation

omplex can consist of any number of production units. Analysis of the work capacities of the major equipment required
for grass plantations indicates that the minimum area for a
plantation production unit is about the same as that for
deciduous plantations with the same productivity.

VI. ECONOMICS OF THE ENERGY PLANTATION[TM]

A.   Establishment and Start-up Costs

To establish an Energy Plantation[TM], costs are incurred to clear and prepare land. Other costs are incurred to
convert land ready for planting into an operating plantation.
These are called start-up costs, and they include the costs
for producing and planting clones. The costs of land clear-
ing and preparation are treated as capital costs incurred in
establishing the plantation. Preparation includes the cost
of spreading fertilizer and lime on the cleared land. In the
case of deciduous plantations, the cost of planting about 5%
of the land to produce clones for planting throughout the
plantation is included in the capital cost.

It is contemplated that as land is cleared, it will be
prepared for planting and then be planted at the first
opportunity when weather, soil moisture, and seasonal factors are conducive to good germination of clones and subsequent plant growth. Thus, during the initial start-up period,
selected portions of the plantation are progressively cleared
and then planted to effect a continuous operating schedule.
When the plantation is fully established, portions of it will
be in all of the different operations simultaneously; some
portions in a growth phase, some in a harvest phase, and
others being regenerated. This continuous operating schedule
promotes a more-or-less constant supply of Energy Plantation[TM] fuel.

For an assumed plantation operation in the south, the
land-clearing and preparation cost was estimated to be about
$40 per acre. Start-up costs were estimated to be about $90
per acre. These costs are believed to be representative for
plantations elsewhere having the same planting density and
harvest schedule.

For the establishment and start-up operations of a grass plantation, about 2 years will probably be needed. The land-clearing and preparation costs are estimated to be about $40 per acre, and the start-up costs about $28 per acre. The start-up costs for a grass plantation are significantly lower, because purchasing and planting grass stolons is much cheaper than producing and planting clones.

B.   Capital and Operating Costs

Estimates have been made of the equipment and manpower requirements and capital and operating costs for plantation production units having minimum practical areas for maximum annual plant-material productivities of 8, 9, and 10 oven-dry tons per acre. These productivities span the estimated range of maximum productivity for Energy Plantation^{TM}. A planting density of 4 ft^2 per clone, and a harvest schedule calling for the first harvest when the stand is a year old, followed by five harvests from stump regrowth with 2-year intervals between them, has been assumed for the estimates. It is assumed for these estimates that the average distance plant material is hauled from where it is harvested to the fuel-using facility is about 11 miles. Of this distance, about 5 miles are inside and 6 are outside the plantation.

Factors involved in these cost estimates include the work capacity and personnel, fuel, supply and maintenance requirements for the equipment used, and the functions performed in the plantation. They also include the capital costs and estimated service lives of the equipment visualized for plantations [6]. These cost estimates are based on 24 hours per day, 7 days per week operation. The estimated rates for the various skills involved and payroll fringe benefits (20% of the pay rate) are based on rates characteristic for large farming operations where much of the work is done by hired hands. The costs and pay rates are generally at levels in effect early in 1975. The manpower required for a minimum plantation production unit of 28,500 acres is estimated to be 86 full-time and 48 part-time people.

The breakdown of the estimates of capital and operating costs for a plantation consisting of four 28,500-acre
production units is shown in Table II. The calculation of
the revenue required to cover all costs involved in growing
and delivering plant material to the fuel-using facility has
been made according to the utility method of financing [12].
In making the calculations for this table, the following
assumptions were used: (a) a 15% after-tax return on equity,
(b) a debt-to-equity ratio of 75 to 25, and (c) an interest
rate of 7.5%. It is assumed that the land in the plantation

TABLE II

Capital and Operating Costs for Plant Material Grown on Energy Plantations[TM] [a]

	Deciduous-Species Plantation		Grass Plantation	
	Thousands $	Percent of total revenue required	Thousands $	Percent of total revenue required
Plantation investment				
1. Machinery and buildings	5,874		4,255	
2. Land clearing and preparation	5,514		4,593	
3. Total plant investment	11,388		8,848	
4. Interest during construction				
a. Machinery and buildings	138		100	
b. Land clearing and preparation	775		431	
5. Start-up	9,924		3,198	
6. Working capital	1,543		1,610	
7. Total capital investment (TCI)	23,768		14,187	
Annual operating costs				
8. Fuels	593	4.2	1,646	12.8
9. Land rental	2,565	17.2	2,565	20.0
10. Payroll	3,337	23.3	3,265	25.4
11. Administration and general overhead	790	5.5	770	6.0
12. Operating supplies	664	4.7	412	3.2
13. Repair parts	1,004	7.0	762	5.9
14. Local taxes and insurance	307	2.2	239	1.0
15. Total annual operating	9,260	64.8	9,658	75.2
Depreciation				
16. Total depreciation	2,895	20.2	1,909	14.9
Returns				
17. Allowable gross return	1,565	11.0	933	7.3
18. Federal income tax	578	4.0	345	2.7
19. Total capital charges (TCC)	5,039	15.0	3,187	10.0
Total revenue required	14,299	100%	12,845	100%
Total revenue required per oven-dry ton of plant material delivered	$13.94		$12.53	

a. Plantation productivity, 9 oven-dry tons per acre per year; annual plant
material produced, 1.03 x 10^6 oven-dry tons; number of plantation units, 4 (114,000
acres).

is rented from its owner, the annual rental being taken as
the value of the land --assumed to be $240 for a specific
site--times the utility rate of return.

It can be seen from Table II that payroll and payroll
overhead costs are the single largest contributor to the
total estimated cost of operating the plantation. These two
costs together account for almost 30% of the total annual
cost. The next two largest cost elements are depreciation
and land rental.

The cost estimates for producing a warm-season grass
on an Energy Plantation^{TM} are also shown in Table II. The
estimated costs of operating a grass plantation are slightly
lower than those for a deciduous species plantation.

The cost breakdown is very similar to the deciduous
species case except for two significant differences. The
fuel requirement for grass plantations is estimated to be
nearly three times that of deciduous species plantations.
The fuel consumed in the harvester-windrowers is almost
entirely responsible for this difference. The other major
difference is the substantially smaller capital cost of machinery and buildings, which leads to a difference in annual
depreciation charges.

C. Cost of Fuel Produced

The cost of the fuel produced on an Energy Planta-
tion^{TM} is simply the total revenue required for the operation
divided by the amount of material produced. For the operations whose cost estimates are shown in Table II, the cost of
the plant material produced is $13.94 per oven-dry ton of
deciduous material and $12.53 per oven-dry ton of grass. At
a useful fuel value of 5700 BTU per air dry pound of plant
material, these costs correspond to an energy cost of $1.22
and $1.10 per million BTU of useful heating value for decid-
uous and grass materials, respectively. The useful heating
fuel value of oven-dry material, which was taken as reference in Table II, is probably higher and close to 6000 BTU
per oven-dry pound. In that case, the cost of energy would
be reduced to $1.16 and $1.04 per million BTU per deciduous
and grass materials, respectively.

Cost estimates were also prepared for plantations with different land productivities, different scales of operations and different harvesting and planting cycles to investigate the sensitivity of the plant-material cost to these factors. Figure 2 summarizes the results of this analysis. It is apparent that the total number of harvests per plantation cycle has the largest beneficial effect on the cost of plant matter. Increasing the total number of harvests from six to eight, while maintaining the sustained productivity and all other factors constant could lower the cost of plant matter by 9% to $12.70 for the case discussed

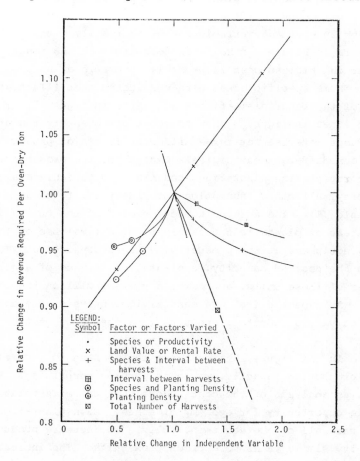

Fig. 2. Generalized effect of variations in factors under plantation management control on revenue required for plant material production based on medium-yield deciduous species plantation (9 oven-dry tons per acre-year).

ˋn Table II. Plant matter productivity or the replacement of
a medium-yielding species by a high-yielding species has a
differential effect similar to that of the number of harvests,
but the range of relative changes in yields is very limited
(from about 9 to 10 tons per acre-year, or 11%). Altering
the planting density or the intervals between harvests only
(in combination or not with a change to more productive
species) has a relatively minor effect on the cost of plant
matter.

Land value or rental rate has a strong detrimental
effect on the cost of plant matter. The combined effect of
the simultaneous modification of two or more operating factors or independent variable cannot be readily estimated from
the data of Figure 2 other than to state that the resulting
impact on biomass cost is probably favorable or not.

Some specific cases were considered that illustrate
some of the beneficial effects suggested in Figure 2. The
solid lines in Figure 3 show the cost per oven-dry ton of
deciduous material for three different levels of productivity
and four different capacities. Except for the productivity
and corresponding plantation area required to achieve a given
capacity, all other assumptions are identical to those used
in Table II. The acreage noted for each of the curves are
the areas of plantations units insuring optimal use of the
heavy equipment at the productivity level considered. To
reach the desired capacity, a plantation is made of a certain
number of these units, approximately 4 per million ton capacity. As expected from the sensitivity analysis, a change in
productivity of about 11% results in a change of cost of
about 5%.

In the figure, the cost per ton is seen to increase
with production capacity, which is perhaps surprising. However, an analysis of the costs shows that there is relatively
little opportunity for economy of scale. Each plantation
unit uses all of the equipment fully. To increase production
capacity simply takes more production units. The increase in
cost with increase in production capacity is the result of
having to transport the material a further distance to the
central fuel-using facility. This characteristic of the cost

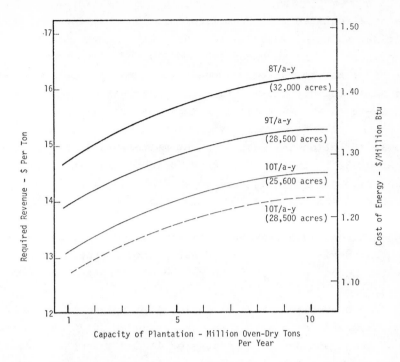

Fig. 3. Effects of productivity and planting density and harvest schedule on revenue required for deciduous plant material. (————) 4 ft^2 per plant; first harvest, 1 year; subsequent harvests, 2-year intervals. (- - -) 8 ft^2 per plant; first harvest, 1 year; subsequent harvests, 3-year intervals.

of Energy Plantation^{TM} fuel must be taken into consideration when the central fuel-using facility is designed to arrive at an optimized total system.

The dashed line in Figure 3 corresponds to a case in which the area per plant has been increased from 4 to 8 ft^2 per plant and the interval between harvests from the stump has been lengthened from 2 to 3 years. If one wants to compare this case to the median case (center solid line, which is the reference case for the sensitivity analysis of Figure 2), one has also to assume that an improved species with a higher productivity has been introduced. The combined effect of these changes in plantation operation is a reduction in cost of about 8% for all production capacities, and

the area required per plantation unit is maintained at its
original value of 28,500 acres per unit. Comparing the
plantation operating under the modified operating conditions
(dashed line) to the corresponding original plantation having
the same productivity--10 tons per acre-year (lower full
line)--it is apparent that, as expected from Figure 2, modi-
fying the plantation density and harvesting cycle has a beneficial effect on the cost of plant matter produced (decrease
of about 3%). On the other hand, slightly larger areas per
plantation unit are required to maintain the production
capacity. This example thus shows that a certain amount of
flexibility exists in the design of Energy Plantations;
depending on local land availability and cost and production
requirements, the planting and harvesting operations will
have to be optimized so as to reduce plant matter production
costs.

      The cost estimates of Table II were made on the basis
of production units having an area of about 28,500 acres made
up of large enough single tracts of land for the time lost in
moving the equipment from one tract to another to be negligible compared to the actual working time. This is also the
hypothesis on which the estimate of land available for plantations was made. As mentioned earlier, if small lots could
be used economically for plant matter production, more areas
of the country could rely on Energy Plantations for their
energy needs. This problem is presented in Figure 4. The
reference point (circled on the figure) corresponds to the
case discussed in Table II. The cost data was generated for
a location in Mississippi for which the land availability
study had indicated that land available for plantation
amounts to about half of the geographical area, i.e., a
28,500 acre unit will be located in a 57,000 acre area. The
other points on the figure correspond to a situation where
the land available for plantation amounts to only 17% of the
geographic area but also where the plantation units are made
up of lots of variable sizes. The assumed land composition
of the units is shown at the bottom of the figure. The
situation is thus highly unfavorable--longer travel distances and waste of time to move equipment from plot to plot--

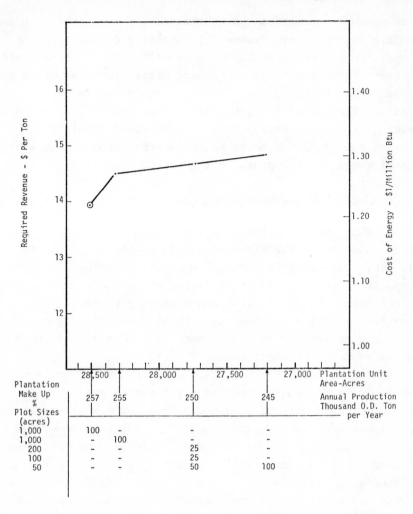

Fig. 4.   Influence of plantation make-up on plant matter production costs.

and is representative of a semi-rural area of nothern Virginia. The harvesting equipment was assumed to be identical to that of the base case, and as a result of the time wasted to move equipment around, the area of a plantation unit and thus the amount of material produced per year is reduced as the size of the individual plots making the plantation decreases. The amount of equipment needed to transport the plant matter to its point of use was adjusted to take

into account increasing travel distances. As can be seen on
the figure, the cost of material increases steadily as the
dispersion of the plantation increaes, but even in the case
of plantations made only of 50 acre plots, the cost of plant
matter is less than $15 per oven-dry ton delivered to a
central point under the productivity conditions considered
here. The cost of energy in this case would be $1.30 per
million BTU of useful heat, which is still competitive with
fossil fuels in many areas of the country.

D. Potential of the Energy Plantation

This analysis indicates that plant matter can be produced at costs ranging from about $13 to about $16 per oven
dry ton. Expressed in terms of useful fuel value, this corresponds to about $1.14 to $1.40 per million BTU, which is
competitive with fossil fuel costs in many parts of the
country. It was also shown that the Energy Plantation concept can be adapted to local conditions with little changes
in its economic attractiveness.

If the 100 million acres of land estimated to be
available were used to generate plant matter fuel, at an
average productivity of 9 oven-dry tons per acre-year, about
10 quadrillion BTU of useful fuel value could be generated
per year, or about 14% of the total national consumption of
energy. This is about twice the amount of energy that could
be recovered from all sources of municipal solid wastes and
forest and agricultural residues. If converted to electricity, the energy potentially available from Energy Plantations would amount to almost 1200 billion kw-hr or about 61%
of the total electricity consumption of the country in
1974 [13].

Obviously one cannot expect all land available to be
converted to Energy Plantations, but the estimates presented
above show that this concept has considerable potential for
becoming a significant new alternative source of energy.

REFERENCES

1. John A. Alich and Robert E. Inman, "Effective Utiliza-
   tion of Solar Energy to Produce Clean Fuel," Report
   No. NSF/RANN/SE/GI 38723/FR/2.
2. Darrel S. Metcalfe, Forage Statistics in "Forages"
   (Maurice E. Heath, Darrel S. Metcalfe, and Robert E.
   Barnes, eds.) Chapter 7. Iowa State University Press,
   Ames, Iowa, 1973.
3. Hugh W. Bennett, Johnsongrass, dallisgrass, and other
   grasses for the humid south in "Forages" (Maurice E.
   Heath, Darrel S. Metcalfe, and Robert E. Barnes, eds.)
   Chapter 32. Iowa State University Press, Ames, Iowa,
   1973.
4. Glen W. Burton, Bermuda grass in "Forages" (Maurice E.
   Heath, Darrel S. Metcalfe and Robert E. Barnes, eds.)
   Chapter 31. Iowa State University Press, Ames, Iowa,
   1973.
5. Ernst Assmann, "The Principles of Forest Yield Study."
   Pergamon, Oxford, London, 1970.
6. Opekasit, Inc., Cincinnati, Ohio (Agricultural engineering consultants to InterTechnology Corporation),
   personal communications.
7. Solar Energy Intelligence Report, July 19, 1976.
8. Todd W. Bowersox, Pennsylvania State College, private
   communication with Jean F. Henry, InterTechnology
   Corporation, June 1974.
9. Poplar Council News, Spring 1975.
10. National Swedish Board for Energy Source Development,
    Note on Short Rotation Forestry, July 1976.
11. Joseph R. Saucier, USDA Forest Service, Athens, Georgia,
    private communication with Jean F. Henry, Inter-
    Technology Corporation, July 1974.
12. Esso Research and Engineering Company, "Description of
    Gas Cost Calculation Methods Being Used by the Synthe-
    tic Gas-Coal Task Force of the FPC National Gas
    Survey," 1972.
13. U.S. Bureau of Census, Statistical Abstract of the
    United States, 1975.

RESULTS OF A NATIONAL SURVEY OF CELLULOSIC WASTE

CHARACTERISTICS AND PROCESSES

FOR THEIR CONVERSION TO ENERGY FORMS*

E. Milton Wilson and John M. Leavens

The Ralph M. Parsons Company
Pasadena, California

I. BACKGROUND

As part of its role in fostering activities that will
lead to environmentally sound waste-disposal practices, the
U.S. Environmental Protection Agency has sponsored a study of
the "Utilization of Wastes as Fossil Fuel Energy Substitutes."
The Ralph M. Parsons Company was the contractor selected to
perform this review and analysis. The program consisted of
four phases:

1. Compilation of information on waste-to-energy technology and on-going developments.

---

* The work reported herein was supported by the
Industrial Environmental Research Laboratory, Cincinnati, of
the U.S. Environmental Protection Agency under Contract
No. 68-02-2101. The Project Officer was Mr. Harry M. Freeman.

2. Establishment of the quantities and characteristics
of the several categories of wastes (exclusive of those from
mining operations) and their projected trends.

3. Engineering analysis (technical, environmental, and
economic) of seven candidate processes selected to represent
a range of input waste characteristics and output energy
forms.

4. Development of recommendations and research and
development needs that would encourage local government and
industry to convert increasing quantities of wastes to
energy.

More than four thousand articles and reports exist
describing various aspects of waste-to-energy technology.
Analagous literature on gasification and liquefaction processes for converting coal to clean fuels greatly extends
this list. The subject matter can most conveniently be cate-
gorized into the areas of (a) combustion, (b) thermochemical
conversion, and (c) biological conversion. Each has its
relative advantages and both research and commercialization
activities are continuing in all fields.

The ranking of a supposed superiority of one individual
system over another was avoided throughout the project, since
the processes are in various stages of development and localized conditions of waste supply and energy needs can seriously affect conclusions. Even a preliminary analysis of the
potential technical and economic feasibility of a proposed
scheme, however, requires considerations of mass and energy
balances. The most serious deficiency in developers' reports
describing small batch experiments or continuous processes
of less than a few kilograms per hour throughout is the
absence of meaningful information that would permit calculation of such balances. This is particularly true for thermochemical conversions based on rather lengthy tests, with
heat being supplied from an external furnace. The time-
temperature history of the sample is poorly defined, and
whether any net energy at all was recovered from the original organic fraction cannot be stated with assurance.

An additional problem encountered in reviewing experimental results concerns the feed composition. True waste, almost without exception with regards to its source, is quite heterogeneous within any given sample and variable over time. Experiments with small quantities of real waste material oftentimes suffer from nonrepresentative samples or particle size distribution not at all similar to what would be used in a full-scale processing plant. Moisture contents can be significantly different from that encountered in later development work. Synthetic samples, while permitting more reproducible conversion results, yield little information as to potential commercial feasibility.

The above-cited problems in evaluating laboratory waste-to-energy research necessitates pilot-plant and semi-works scale-up testing before the value of the process can be fairly judged. Other than in a few well-proven technologies, sufficient new problems have been encountered in the additional testing to justify the continuing need for it. Costs of development through the 200 mg/day level can be in excess of $10 million, and hence, there are only a limited number of candidate types that will be constructed as commercial plants in the next 4- to 5-year period.

II. WASTE-TO-ENERGY SYSTEMS

Except in the case of rather limited quantities of specialized liquid industrial wastes, the value of waste materials for fuels is essentially due to their cellulosic fraction. This is true whether we are dealing with logging residuals, feedlot manure, or municipal solid waste (MSW). The energy inherent in the cellulose is ultimately obtained by complete oxidation of the carbon and hydrogen. It is the variation in the steps leading up to that final conversion to carbon dioxide and water that results in the number of distinctive processes under evaluation.

The variations are designed to permit best utilization of heat evolution by energy consumers having different needs. The primary consumer demand is usually one for a particular physical form; i.e., he has a preference for a solid, liquid, or gaseous fuel. This can be a result of historical

practice, process requirements, ownership of existing equipment, transportability considerations, or emission control
costs. The fuel need could be for use in any of a number of
types of furnaces/steam generators, cement kilns, gas tur-
bines, internal combustion engines, or home kitchen stoves.
Final application can range from process heating or cooling,
electric power generation, or transportation. Initial properties of the waste can also influence the choice of the conversion process; a high moisture content, e.g., would suggest
a process whereby energy would not be consumed in evaporating
water, such as biological anaerobic digestion to methane.

Only in the case of direct combustion of the cellulose
is the oxidation step accomplished in one step. In the other
cases, conversion of the cellulose to another chemical form
is accomplished and it is this form that is later oxidized.
Table I shows the several processing alternatives, both with
different initial waste and final fuel forms.

III.   UNITED STATES ENERGY REQUIREMENTS

To ascertain the relative position that wastes can
assume in United States energy production, it is necessary to
examine some limited statistics on present and anticipated
patterns. The subject is of course a highly complex one,
involving interrelated dynamics of national and international
economics, politics, environmental considerations, research
and development policies, and sociological habits. Figure 1
shows United States energy consumption by sector, while
Figure 2 shows it by generating source. While waste-to-
energy conversion systems will never attain any high percentage of the latter values, every contribution to these totals
is important because of the absolute magnitude of the costs
involved. Even under quite conservative assumptions, waste
materials could soon contribute energy equivalent to several
billion dollars per year of foreign petroleum.

TABLE I

Waste-to-Energy Conversion Matrix

Waste form	Using or derived form		
	Solid	Liquid	Gas
Solid	Direct grate combustion	Flash pyrolysis	Pyrolysis
	Suspension firing of organic fraction	Fermentation	Anaerobic digestion
	Pelletized organic fraction	Reaction with carbon monoxide or hydrogen	Hydrogasification
	Low-temperature pyrolysis		Partial oxidation
	Feedstock for chemical conversion		Water gas reaction
Liquid	Incorporation into porous solid fuel	Liquid incinerator + waste-heat boiler	Direct pyrolysis or additive to solid waste pyrolyzer
			Anaerobic digestion
Gas	While solid hydrocarbons could be synthesized, there are no practical applications to date	Conversion to methanol or liquid hydrocarbons	Combustor + waste-heat boiler
			Conversion to methane or ammonia

207

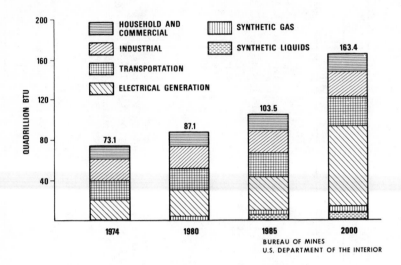

Fig. 1.   United States energy consumption by sector,
1974-2000.   From Bureau of Mines, U.S. Department of the
Interior.

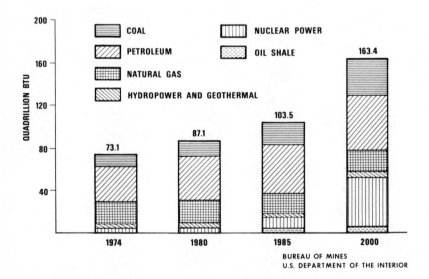

Fig. 2.   United States energy consumption by source,
1974-2000.   From Bureau of Mines, U.S. Department of the
Interior.

IV. QUANTITIES OF UNITED STATES WASTE

Estimating the quantities of United States waste that might be economically converted to energy is hampered by several factors. Among the impediments are the following: inadequate records, results from widely varying surveys, determination of the percentage of the totally generated waste of a given type that can be readily collected and transported to a processing facility, and even by an acceptable definition of "waste." Yesterday's residuals are becoming today's by-products; or they are frequently used immediately, internally, for their energy value. This is particularly true in the case of the agricultural and silvicultural industries. Unadjusted apparent waste quantities can be highly misleading, and trends for the amounts truly available probably should have a negative slope. This conclusion could be modified in the cases of deliberately grown biomass which should not be considered a waste material, or if additional research indicates most field crop residuals are not required for proper soil physics and can be collected at reasonable costs.

Table II shows a summary of generally accepted quantities of United States wastes. Such values should be carefully reviewed to note whether they are reported on a water- and ash-free basis, a correction that can lead to as much as a 3:1 error in conclusions formed about the total quantity of energy available. An additional serious error can be introduced by the overly simplistic mathematics of factoring the above 567.6 million tons by a heat of combustion of, say, 7500 BTU/lb (17.44 mj/kg) and claiming that 8.65×10^{15} BTUs are therefore available, "equivalent to 1.37 billion barrels of oil." While a mathematical equivalency is valid, in actual practice, as much as 50% of the waste energy could be nonrecoverable because of conversion losses and boiler efficiencies. It is essential that all energy calculations be made with actual cases of a given waste through final products of a specific processing system.

Several studies [1-4] have established major crop residue factors, usually defined as the above-ground portions

TABLE II

United States Waste Quantities Discarded[a]
(Millions of combustible dry tons)

Source	Quantity
Urban	
Household and municipal wastes	84.2
Sewage solids	6.9
Commercial wastes	31.0
Manufacturing plant wastes	11.8
Demolition residues	3.8
Manufacturing and processing	
Wood-related wastes	25.7
Textile and fabric wastes	0.3
Nonfabric synthetic wastes	0.4
Food processing solids	0.8
Miscellaneous	0.1
Agricultural	
Animal wastes	206.7
Crop wastes	170.0
Forest and logging residues	25.9
Total	567.6

a.   Source: International Research and Technology
Corp.

of the crop plant that are not harvested. In terms of tons
of dry residue per acre of harvest, typical values are 2 to 4
for corn, 0.7 to 1.6 for wheat, 1.2 to 1.5 for cotton, and
11 to 26 for sugarcane. The corn, wheat, and soybean crops
produce two-thirds to three-quarters of the agricultural
crop waste total.

The quantity of approximately 26 million tons of logging residue is typical of several investigators [1, 2, 5]
who assumed wood under 4 inches in diameter and nonsalable
trees would not be available. Ellis [6], recognizing new

wood harvesting technology might permit total recovery of residues in harvested areas, claims that a value of 157 million tons per year would be reasonable. An estimate made by the U.S. Forest Service [7] goes even further and establishes a value of 1154 million dry tons for noncommercial timber, forest harvest residues, other removals, and unused primary manufacturing residues; the first entry, estimated at 985 million tons, is basically a one-time inventory, not an annually recurring quantity.

Because of the need for adequate disposal of household refuse, more is known on the characteristics of MSW than any other waste stream. Table III shows quantities of the various components of it discarded in 1971.

TABLE III
Municipal Waste Generation in 1971[a]

Materials	Million tons	Percent
Paper	39.1	31.8
Glass	12.1	9.7
Metals	11.9	9.5
Plastics	4.2	3.4
Rubber and leather	3.3	2.6
Textiles	1.8	1.4
Wood	4.6	3.7
Product totals	77.1	61.7
Food wastes	22.0	17.6
Yard wastes	24.1	19.3
Miscellaneous inorganics	1.8	1.4
Total	125.0	100.0

a. These figures are on an "as generated" basis; that is, the product categories are considered air-dry, and moisture accumulated in storage or transport was not considered. Source: Office of Solid Waste Management Program.

The percentage composition is estimated in Table IV. Reports giving detailed information on the characteristics, quantities, and trends of MSW are cited as references [8-12].

TABLE IV.
Composition of Municipal Waste

Component	Wt % as received
Paper products	44.91
Plastics	4.09
Textiles	3.16
Wood	2.23
Yard trimmings	2.06
Food wastes	10.94
Rubber and leather	1.22
Fines (<1 in)	11.45
Metal	9.89
Glass and ceramics	10.05
Total	100.00

V.   CANDIDATE SYSTEMS

The candidate systems analyzed were as follows:

1.   Water wall incineration, as exemplified by the new 1200 TPD RESCO plant at Saugus, Massachusetts, and the older 600 TPD facility at Hamilton, Ontario, Canada.

2.   An urban central heating-cooling plant fueled with solid waste (the 720 TPD Nashville Thermal Transfer Corporation).

3.   A supplementary refuse derived fuel (solid) plant, using the new 2000 TPD City of Chicago-Commonwealth Edison plant as the example.

4.   The air-blown slagging pyrolysis system of Andco-TORRAX as a steam generator.

5. The oxygen-blown slagging pyrolysis system of Union
Carbide Corporation-PUROX as a source of medium heating value
utility gas and as a synthesis gas for conversion to methane,
methanol, and ammonia.

6. The flash pyrolysis waste-to-liquid fuel system of
the Occidental Research Corporation.

7. The 200 TPD mobile agricultural waste pyrolysis
system (to solid fuel) of the Georgia Tech Engineering Experiment Station.

Each candidate has progressed to the stage of development (or in Case 1, the commercialization) where serious
interest is being shown in their specialty fuel or energy
output. Start-up problems have typically been encountered,
but these have been no different than those experienced in
the chemical industry during scale-up testing. Any initial
environmental emissions in excess of federal or state standards have been found to be controllable within the existing
technology. The ability of the plants to compete with fossil
fuel costs must await full-scale plant evaluation in many
cases. Plant economics are strongly affected by local
conditions; the marketability of recovered materials such as
magnetic metals, aluminum, and glass; and the costs of alternative means of waste disposal.

A discussion of the features of these systems cannot be
presented within the limitations of this review. References
[13-19] contain most of the thus-far released information on
process descriptions, efficiencies, and fuel characteristics.

VI. CONCLUSIONS

Great progress is being made in developing suitable
technology for converting cellulosic waste materials to
energy. While simultaneously reducing the burden of the
normal high costs of disposal of these materials, significant
supplanting of fossil fuel needs can be accomplished.

In the "Third Report to Congress--Resource Recovery and
Waste Reduction" [8], EPA estimated that there exists in the
nation's most densely populated areas alone enough converti-
ble municipal wastes to provide 1085 trillion BTU of energy

by 1980, or 1.3% of the amount of energy forecast by the
Bureau of Mines to be used in that year.   This is a significant amount of energy.   For example, it is equivalent to 37%
of the projected input from the Alaskan pipeline for 1980.
It would supply more than the required energy for combined
residential cooking and clothes-drying needs or 85% of the
energy demand of all commercial air conditioning.

Another recent EPA study concludes that, even if it is
assumed that no conversion plant smaller than 100 tons/day
capacity can be justified economically, enough energy could
be recovered from the normal solid waste stream to supply up
to 3% of the nation's total energy requirement, or enough
energy to provide for all of the country's residential and
commercial lighting needs.   Once a clearer perspective emerges
on the true availability of the potentially vast quantities
of forestry and agricultural wastes in the United States,
these figures could perhaps double or triple.

The conversion of wastes into energy is no longer in
the theoretical stages of development, but has moved into the
beginning phases of commercial application.   Based on energy
recovery systems existing or planned at the present time, it
is projected in the EPA report to the Congress that by 1980
almost thirty cities and counties around the country should
be operating the equivalent of thirty-six 1000 tons/day
plants, recovering an estimated 85 trillion BTU/year.

The report summarized here will be published early in
1977 by EPA.   Government and industrial organizations desir-
ing to ascertain what position they should take in considering the incorporation of a waste-to-energy plant into their
waste management plans should secure a copy through the usual
U.S. Government publishing sources.

REFERENCES

1. International Research and Technology Corporation (IR&T),
   "Problems and Opportunities in Management of Combustible Solid Waste." Prepared for the National Environmental Research Center, EPA, October, 1972.
2. A. Poole, The potential for energy recovery from organic
   wastes, in "The Energy Conservation Papers" (R. H.
   Williams, ed.). Ballinger Pub. Co., 1975.
3. J. Knutson, G. E. Miller, and V. P. Osterli, "Crop
   Residues in California--Some Factors Affecting Utilization." University of California, Davis, Division of
   Agricultural Sciences Leaflet 2872, February 1976.
4. J. A. Alich and R. E. Inman, "Energy from Agriculture."
   Reprinted from the Tenth Intersociety Energy Conversion
   Engineering Conference (IECEC), 1975.
5. J. B. Grantham and T. H. Ellis, Potentials of wood for
   producing energy, J. Forestry, September 1974.
6. T. H. Ellis, The Role of Wood Residue in the National
   Energy Picture in "Wood Residue as an Energy Source."
   Forest Products Research Society, Proceedings No.
   P-75-13, pp. 17-20 (1975).
7. "The Feasibility of Using Forest Residues for Energy and
   Chemicals," National Science Foundation by USDA Forest
   Service, Report NSF-RA-760013, March, 1976.
8. "Third Report to Congress, Resource Recovery and Waste
   Reduction," (SW-161), Office of Solid Waste Management
   Programs, U.S. Environmental Protection Agency, 1975.
9. F. A. Smith, "Comparative Estimates of Post-Consumer
   Solid Waste," U.S. EPA/530/SW-148, May 1975.
10. W. R. Niessen and S. H. Chansky, "The Nature of Refuse,"
    Proceedings 1970 Nat. Incinerator Conf., pp. 1-24.
    ASME, New York, 1970.
11. W. R. Niessen and A. F. Alsobrook, "Municipal and
    Industrial Refuse: Composition and Rates," Proc. 1972
    Nat. Incinerator Conf., pp. 319-337. ASME, New York,
    (1972).
12. Base Line Forecasts of Resource Recovery, 1972 to 1990,
    Midwest Res. Inst. Rep. 3736-D. OSWMP, U.S. EPA,
    March, 1975.

13. G. L. Huffman, "EPA's Program in Environmental Research in Wastes-As-Fuels," for presentation at the Inst. Gas Technol. Symp. on "Clean Fuels from Biomass, Sewage, Urban Refuse and Agricultural Wastes," Orlando, Florida, January 29, 1976.

14. R. G. Schwieger, "Power from Waste" Power Mag. February 1975.

15. N. W. Snyder, "Energy Recovery and Resource Recycling," Chem. Eng., October 21, 1974.

16. E. M. Wilson and H. M. Freeman, "Processing Energy From Wastes," Environ. Sci. & Technol. 10(5), 430-435 (1976).

17. H. W. Schulz et al, "Resource Recovery Technology for Urban Decision-Makers," Urban Technol. Center, Columbia Univ., January 1976.

18. L. Van Poolen (ed.), "Energy Recovery from Solid Waste." NASA/ASEE Systems Design Institute, Univ. of Houston, Coll. Eng., September 1974.

19. Bechtel Corporation, "Fuels from Municipal Refuse for Utilities: Technology Assessment." Prepared for Electric Power Res. Inst., Palo Alto, California, March 1975.

ELECTRICAL ENERGY STORAGE CAPACITY

OF LIGNOCELLULOSE PYROLYZATES

Alan Wilkinson and Cornelius Steelink

Department of Chemistry

University of Arizona

Tucson, Arizona

Waste products from the lumber and paper/pulp industries accumulate at the rate of millions of tons annually in
the United States. Sawdust, newsprint, solid municipal
wastes and pulp effluent are the major constituents of this
accumulation. Many proposals have been developed to utilize
the fuel potential of these materials. We have focused our
attention on these lignocellulosic materials as potential
sources of stored electrical energy.

If these materials could be converted to conducting,
electroactive solids by inexpensive means, they could be
fashioned into electrodes for large primary or secondary batteries. These batteries would be low-energy-density devices
(compared to fuel cells or molten metal cells). However,
since the lignocellulose materials are cheap and abundant,
and since size and portability are not critical to stationary

operations, the energy density need not be the critical parameter. Is there a potential market for low-energy cells?
One possible use of a low-cost rechargeable cell could be for
electrical utility load leveling. Another use would be the
storage of electrical energy produced from solar energy collectors. Such a battery complement to solar collectors was
suggested some time ago by Farrington Daniels [1]. Our preliminary results [2] with pyrolyzed lignin electrodes have
prompted us to study the properties and potential of lignocellulose batteries.

The two principal constituents of waste wood products
are lignin and cellulose. Industrial lignins are redox
polymers [3, 4] and, therefore, are theoretically capable of
storing or generating electrical energy [5, 6] (see Fig. 1
and 2). Kraft lignin has approximately 6 meq/gm redox

lignin could be transformed into a conducting solid, it could
well function as an electrode in a low-cost battery.

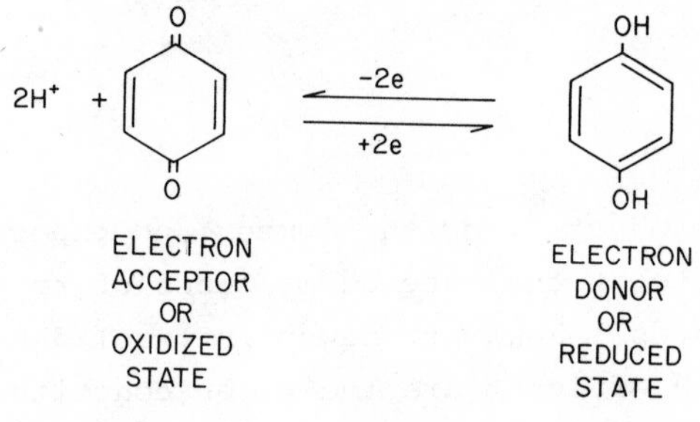

$$2H^+ \; + \quad \xrightleftharpoons[+2e]{-2e}$$

ELECTRON
ACCEPTOR
OR
OXIDIZED
STATE

ELECTRON
DONOR
OR
REDUCED
STATE

Fig. 1.   Electron transfer in Kraft lignin.

I.   RESULTS

During the past two years, we have examined a number
of lignocellulosic materials as electrodes in electrochemical
cells. When Kraft lignin is pyrolyzed at above 700°C under
nitrogen, it forms a cohesive conducting solid. This pyrolyzate retains considerable functionality (Table I). It acts
as a cathode in a primary Zn-lignin cell (Table II). In

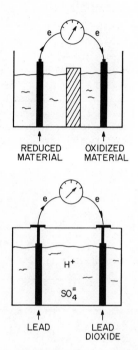

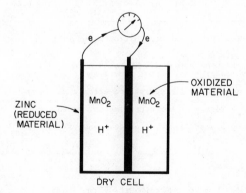

Fig. 2. Typical batteries.

TABLE I

Properties of Modified Lignin Pyrolyzates

| Lignin | Elemental Analysis (%) | | | | | |
Sample	C	H	N	S	Metal	Resistance[a]
A	85.23	1.22	--	--	--	170
B	79.15	1.37	6.55	--	--	155
C	69.09	0.71	6.18	4.43	--	88
D	71.44	0.89	4.01	--	10.31	16

a. Resistance measured across a cylinder of powder, 1 cm x 0.2 cm^2, packed under 1.0 lb pressure. Powder to pass 80 mess screen or less. Sample A: Pyrolyzed softwood Kraft lignin; Sample B: pyrolyzed nitrated softwood lignin; Sample C: Sample B treated with NH_4S_x; Sample D: Sample B treated with $Cr(NO_3)_3$.

TABLE II

Electrochemical Properties of Zinc-Lignin Cell

Lignin[a] Sample	V_{oc}[b]	V_L[c]	V_L[d]
Softwood Kraft[e]	1.22	1.00-0.53	0.91-0.46
Tomlinite[f]	1.24	0.50-0.17	0.44-0.10
Nitrated softwood Kraft	1.39	1.06-0.73	1.04-0.68
Hardwood Kraft[e]	1.24	1.04-0.53	1.06-0.50

a. All samples pyrolyzed under nitrogen for 1 hr at 700°C.
b. Open circuit voltage.
c. Voltage measured across 100 Ω resistance. First number is initial voltage; final number is voltage after 2 hr discharge under load.
d. Voltage across 100 Ω resistance after a recovery period of 12 hr.
e. Donated by Westvaco, North Charleston, South Carolina.
f. Donated by Domtar Research Center, Senneville, Quebec.

addition, it is slowly regenerated on standing in air; i.e., it slowly catalyzes the reduction of oxygen. This pyrolyzate is easily nitrated; the nitrolignin pyrolyzate has enhanced voltage properties. Other materials that yielded pyrolyzates with similar electrochemical properties were particle, whole wood, and Kraft newsprint mixtures.

Zinc-carbonaceous batteries have been previously
reported. The Air Cell manufactured by Union Carbide [9] is
the most common commercial example; it utilizes a noble-
metal-impregnated carbon black cathode (see Fig. 3). Experimental zinc-quinone cells have also been constructed [10].
Both cathode materials are expensive.

The configurations described above (e.g., metal-carbon
batteries) represent primary cells. These cannot be electrically regenerated; eventually, the electrodes have to be
replaced. Another battery configuration is the secondary
cell, or storage battery. In this assembly, the electrode
material stores electrical charge and releases it on demand.
If anode and cathode were made from pyrolyzed lignin, could
they be reversibly charged? To answer this question,

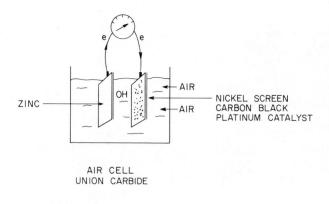

AIR CELL
UNION CARBIDE

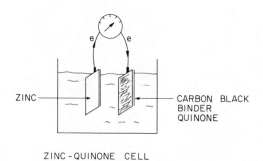

ZINC-QUINONE CELL
SANDSTEDE

Fig. 3.   Carbonaceous batteries.

we examined lignocellulose pyrolyzates as potential anodes.

Kraft lignin pyrolyzates were found to function remarkably well as anodes if they were charged electrically. These pyrolyzate blocks retained a negative charge of -1.35 V versus a Ag/AgCl electrode. Particle board pyrolyzates also performed as anodes (Table III). When Kraft newsprint (25-75 wt %) pyrolyzates were used as anodes, they also retained a negative charge (Table IV).

TABLE III

Discharge Properties of Pyrolyzate Anode

Voltage	Discharge time (min)
1.28	0
1.08	10
0.89	20
0.70	30
0.50	40
0.41	50
0.34	60
0.28	70
0.24	80

Note: 1.0 gm block of pyrolyzed particle board nitrated subsequent to pyrolysis. The block was reduced at -1.40 V (versus Ag/AgCl) for 60 min in 6 N NaOH. It was discharged through 10 Ω versus 0.00 V (Ag/AgCl).

TABLE IV

Electrochemical Properties of Pyrolyzate Anodes

Anode	Cathode	Electrolyte	Energy density (W=min/gm)	Initial voltage	Final Voltage[a]
Newsprint--nitrated	b.	NaOH	2.5	1.05	0.45
Newsprint--untreated	b.	NaOH	0.8	0.9	0.20
Particle board--nitrated	Particle board-- nitrated	Na_2CO_3	0.34	0.82	0.30
Newsprint--nitrated	Newsprint-- nitrated	H_2SO_4	2.70	1.20	0.30
Newsprint--nitrated	PbO_2	H_2SO_4	5.0	1.74	0.80
Newsprint--nitrated after fourth recharge	PbO_2	H_2SO_4	8.5	1.85	1.0
Particle board--nitrated	PbO_2	H_2SO_4	5.6	1.9	0.8

a. After 30 min discharge through 10 Ω .

b. Versus ag/AgCl reference electrode.

Finally, electrochemical cells containing lignocellulose pyrolyzate anode and cathodes were constructed. These
performed as secondary cells (Table IV). Thus, a battery constructed from Kraft newsprint pyrolyzates gave an energy density of 45 W-hr/kg per electrode. This is comparable to the
best lead acid batteries, which have an energy density of
76 W-hr/kg/plate. The cell efficiency (total charge applied/
total charged drained x 100) was 70%.

However, polarization of the pyrolyzate cathodes
caused an unacceptable drop in voltage during the 30 min
discharge period. We have not yet identified the source of
this problem. It does appear to be a cathodic problem, since
substitution of PbO_2 plates for pyrolyzate plates markedly
improved the voltage and energy density of the cell (Table IV).
In fact, a small, fully charged lignocellulose-PhO_2 cell has
operated intermittently in our laboratory for the past four
months (see Fig. 4).

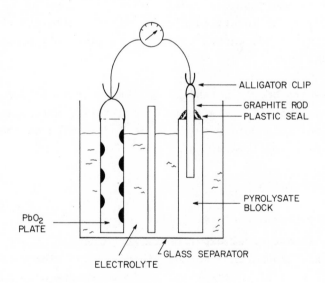

Fig. 4.   Lead dioxide pyrolyzate cell.

II. CONCLUSIONS

We have established that pyrolyzed lignocellulose can
retain an electrical charge. It is more stable in the
reduced state than in the oxidized state. Simple experimental cells approach commercial lead-acid storage batteries in
energy density properties. But, in order to improve the
battery performance, one must know more about the electroactive groups in the pyrolyzates and how they can be modified.
The identification of these oxygen-containing functional
groups is not an easy task, but methodology has been developed by investigors of carbon black and graphite surfaces
[11-17]. Redox functionality in carbon pyropolymers can be
changed by controlled pyrolysis [15, 18]. Thus, our studies
are now focused on three areas:

1. The nature and population density of the electroactive species in the pyrolyzates.

2. The maximization of such species by controlled
pyrolysis and chemical modification.

3. The identification of other parameters that may
affect electrode capability, such as electrolyte, pH, and
permeability.

The results should establish optimum conditions for producing
low-cost storage cells from lignocellulose materials. In
addition, it should provide basic information on the nature
of these pyropolymers.

III. EXPERIMENTS

A.   Preparation of Pyrolyzate

In a typical experiment, 20 gm newsprint and 800 ml
water are ground in a Waring Blendor. The slurry is filtered
and dried in air. The resulting mat is soaked in 10% alkaline solution of Kraft lignin; the wet mats are dried in air
and then heated at 100°C to yield stiff plates. These stiff
mats are stacked between steel plates and baked in a closed
firebrick box at 500°C. Charcoal is packed around the
plates, and nitrogen gas is fed into the kiln during the

baking to maintain a reducing atmosphere. The pyrolyzate plates are now removed from the kiln, restacked, and heated at 1000°C.

B. Evaluation of Pyrolyzate Anode

The pyrolyzate sample is placed in an evaluation cell that contains an Ag/AgCl reference electrode and a working electrode (usually a platinum wire). The pyrolyzate is charged at 10 ma to a voltage of -1.40 (versus Ag/AgCl) for a fixed period of time. After charging, the pyrolyzate is discharged through a known resistance (usually 10 Ω) for a fixed period of time. A voltage-time discharge curve is recorded. From this curve, E^2t/R is calculated, which gives the total electrical energy released during the discharge period. The energy density (watt-hours/kilogram) is obtained from the above value by dividing by the weight of the electrode.

ACKNOWLEDGEMENTS

We wish to acknowledge generous support from the Arizona Foundation and the National Science Foundation.

REFERENCES

1. Farrington Daniels in "Introduction to the Use of Solar Energy" (A. M. Zren and D. R. Erway, eds.) McGraw-Hill, New York, 1963.
2. C. Steelink and Alan Wilkinson, TAPPI 58 (11), 146 (1975).
3. C. Steelink, Advan. Chem. 59, 51 (1966).
4. C. Steelink and J. D. Fitzpatrick, J. Org. Chem. 37,(1972)
5. H. G. Cassidy and K. A. Kun, "Oxidation-Reduction Polymers," Wiley (Interscience), New York, 1965.
6. C. K. Morehouse and R. Glicksman, Electrochem. 106, (1959)
7. K. V. Sarkanen and C. H. Ludwig, "Lignins: Occurrence, Formation, Structure and Reactions," Wiley (Interscience), New York, 1971.
8. B. D. Bogomolev et al., Khim, Drev (Riga), No. 3, 52 (1975).

9. D. H. Collins (ed.), "Primary Power Sources #3," Oriel, London, 1971.

10. G. Sandstede, ed., "From Electrocatalysis to Fuel Cells," p. 333. Univ. Washington Press, Seattle, 1972.

11. H. Binder, A. Kohling and G. Sandstede, Ber. Bunsenges Phys. Chem. 80, 66 (1976).

12. R. E. Panzer and P. J. Elving, Electrochim. Acta 20, 635 (1975).

13. M. L. Studebaker and R. W. Rinehart, Sr., Rubber Chem. Technol. 45, 106 (1972).

14. H. Marsh, A. D. Foord, J. S. Mattson, J. M. Thomas, and E. L. Evans, J. Colloid Interfacial Sci. 49, 368 (1974).

15. V. A. Garten and D. E. Weiss, Rev. Pure Appl. Chem. 7, 69 (1957).

16. H. P. Boehm, E. Diehl, W. Heck and R. Sappok, Angew Chem. Int. Ed. 3, 669 (1964).

17. K. F. Blurton, Electrochim. Acta 18, 869 (1973).

18. Presented at the 52nd Annual Meeting of the Southwestern and Rocky Mountains Division of the American Association for the Advancement of Science, Tucson, Arizona, May 1, 1976 by Alan Wilkinson and C. Steelink.

INDIRECT PYROLYSIS OF KRAFT BLACK LIQUORS

D. W. Goheen, J. V. Orle, and R. P. Wither

Crown Zellerbach Corporation
Camas, Washington

In recent years, the most important pulping system for liberation of cellulose fibers from wood has been the Kraft process. A major portion of all the cellulose pulp produced in the world is now made by this procedure, and, for example, in 1974 in the United States, out of some 36 million tons of chemical pulp, more than 32 million tons were produced by Kraft pulping [1]. The liquors that result from the process contain large amounts of dissolved wood components. For each ton of pulp, approximately 1.6 tons of black liquor solids are produced. These solids are composed of about 60% organic matter, so that some 31 million tons of organic material in addition to cellulose pulp is made by the Kraft industry in the United States.

An important advantage of the Kraft process over other pulping methods involves the comparatively efficient recovery

of inorganic pulping chemicals from the spent pulping liquor.
The chemicals are obtained by burning liquor concentrates in
large recovery furnaces. Pyrolysis of the organic matter
takes place during the chemical recovery operation, but the
pyrolysis products are not separated and have been utilized
only for their fuel values as part of the combustion process
in the recovery operation.

Although it has been recognized that the organic matter of the liquor solids is potentially a rich source of
organic chemicals, attempts to isolate individual chemical
species have met with only very limited success, owing primarily to the diversity in the number of components of the
liquor, the need to convert sodium salts to their free acid
form, and the difficulty in separating soluble components
from large quantities of water.

Dimethyl sulfide and methylmercaptan have been
obtained in commercial quantities by a high-temperature,
high-pressure process in pressure vessels [2]. Other studies
have indicated the possibility of isolation of commercial
quantities of mixed phenolic and neutral products by even
higher temperature and pressure treatments of liquor solids
with alkaline reagents, particularly sodium sulfide [3].
Also of potential industrial importance is the hydrogenation
of isolated lignin using catalysts and pressure to produce
phenol and alkyl substituted phenols [4].

All of these procedures require high pressures with
consequent high capital investments. Thus, it was considered that an investigation of the pyrolytic process that
takes place at atmospheric pressure in recovery furnaces
should be undertaken to determine whether or not useful quantities of products could be isolated by a relatively simple
thermal procedure. Indirect pyrolysis was chosen so that the
products could be isolated and the process would simulate the
pyrolysis occurring in recovery furnaces prior to the actual
combustion of the pyrolytic products. It was felt that even
a relatively low yield of any product might be interesting
and could lead to a comparatively high production rate, since
so much organic material is available in Kraft liquors.

I. GUAIACOL FORMATION

An initial and very simple experiment was performed to
determine the effect of indirect pyrolysis of black liquor
solids. A small quantity of concentrated black liquor
(approximately 10 gm of 50% solids concentration) from
Douglas fir pulping was placed in the bottom of a 22 mm x 175
mm test tube. The tube was carefully heated to expel water.
Following this, the tube containing the dried solids was bent
into an L-shape, simulating a retort. The end containing the
liquor solids was immersed in a Woods' metal bath heated to
about 400°C. On heating, the solids foamed and decomposed,
and an oil collected in the clear portion of the bent tube.
This oil was injected into a gas chromatograph equipped with
a Celanese ester No. 9 column and was found to contain a surprisingly high percentage of guaiacol. In fact, guaiacol was
by far the most plentiful single compound, as shown in
Figure 1.

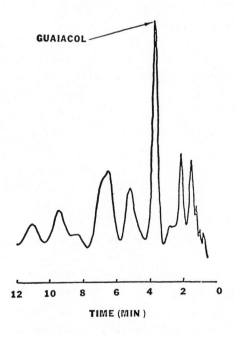

Fig. 1. Guaiacol production.

II.   PYROLYSIS APPARATUS

Based on this initial promising experiment, an apparatus capable of drying black liquor and then subjecting the solids to pyrolytic temperatures was designed and built with the objective of maximizing guaiacol formation. When dried by heat, Kraft black liquor sets into a brittle foam having a volume several times that of the original liquor and forming an insulating barrier to heat transfer. As a result, pyrolysis in stationary equipment is both cumbersome and inefficient. The dry foam can be broken up fairly easily into small particles by crushing. Based on the premise that a device to overcome the foaming tendency of the solids was required, an apparatus (shown schematically in Fig. 2) was designed and built to carry out the pyrolysis experiments. The basic requirement of a crushing device was met by incorporation of a screw inside a pipe. The screw not only served to crush the foam mass, but also carried the solids into the pyrolysis zone, stirred them for better heat transfer and removed the pyrolysis residue.

The screw was made by welding together two steel 1 ft long, 1 inch diameter wood bits and the tubing was fabricated from 1 inch diameter pipe. The screw was turned by a Vicker's drive at 30 rpm and "floated" in the pipe with no support bearings. As shown in Figure 2, the distillate could be taken off in either direction.

Two modifications of the equipment were used initially. In one, the pipe was composed of stainless steel and was heated by means of a fused salt bath in a stainless steel pan built around the middle section of the pipe. In the other, the pipe was composed of black iron and was heated by an electric tube furnace placed around the middle of the pipe.

Superheated steam was introduced in countercurrent flow in a number of pyrolyses. Superheating was done in two stages with a flame-heated Fisher laboratory superheater connected in series with a 1/4 inch stainless steel pipe furnace.

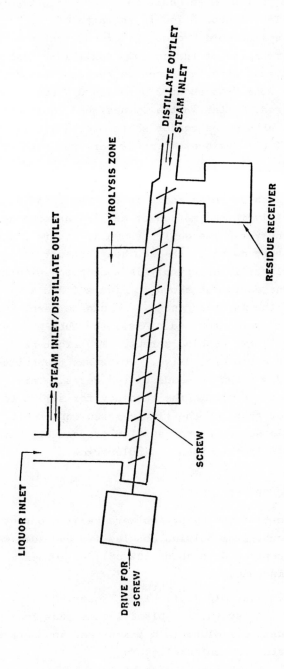

Fig. 2.   Initial pyrolysis reactor.

Introduction of black liquor into the system was by dropwise fall from an addition funnel onto the rotating screw. The liquor dried in the area of the liquor inlet, and the solids were broken up and carried into the pyrolysis zone by rotation of the screw. Auxilliary heat, provided by an electrical tape, was used to keep the temperature near the addition zone from about 210 to 300°C. Some trouble was encountered, in the initially designed equipment, with solid build-up around the center shaft of the screw. However, a number of pyrolysis experiments were run with this type of equipment.

Other variations in the design were devised and studied such as those shown in Figures 3 and 4. They differed mainly in the method of heating and the design of the screw conveyor. The design shown in Figure 3 was not very satisfactory owing to the low heat capacity of superheated steam. Insufficient pyrolysis took place with this equipment. The equipment shown in Figure 4 had several variations of the initial design. As can be seen, two liquor addition ports were provided. Also, the tube pyrolyzer was fitted with two meshing screws. The right-handed screw was turned by a Vicker's drive. Power was transmitted to the left-handed screw by means of two gears. This equipment was designed to overcome the tendency for solids to build up on the center shaft of the single-screw equipment. In practice, the double screw was not able to prevent all build-up, but was more effective than a single screw.

III. LIQUOR PYROLYSIS

Most of the pyrolyses were carried out using fresh black liquor from pulping Douglas fir and some western hemlock concentrated to about 50% solids. Other materials studied included:

1. Year-old Kraft black liquor.

2. A "synthetic" black liquor made from precipitated Kraft lignin dissolved in a mixture of appropriate chemicals to simulate Kraft black liquor.

3. Black liquor with added sulfur and/or sodium sulfide.

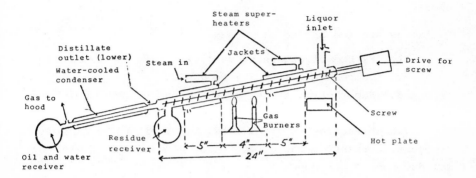

Fig. 3.   Second indirect pyrolysis system.

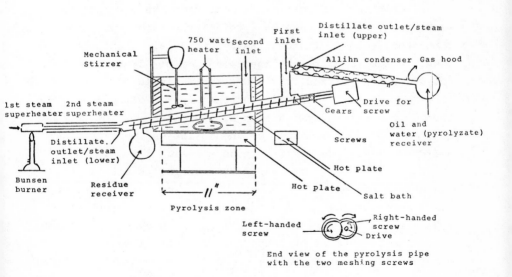

Fig. 4.   Final indirect pyrolysis system.

    4. Black liquor with added sodium carbonate.
    5. Soda black liquor.
    6. Sodium-base sulfite spent liquor.
    7. Calcium-base sulfite spent liquor.
    8. Kraft black liquor from pulping southern hard-
woods.

Variables studied included:

    1. Method of attaining pyrolysis temperature.
    2. Temperature from 325 to 500°C.
    3. Superheated steam for sweeping out pyrolytic
products.
    4. Direction of steam flow, i.e., either concurrent
or countercurrent to the direction of flow of the liquor
solids.
    5. Steam-flow velocity.
    6. Single or double screw.
    7. Rotational speed of screw.
    8. Black liquor addition site.

The tabular results of the various experiments are shown in
Tables I-V.

    Choosing the best conditions from the experimental
work, two runs were made on a relatively large, but still
laboratory scale. In the first, a material balance study was
made. A sample of black liquor, 656 gm of 53% or 348 gm of
solids, was pyrolyzed at 415°C using a single-screw tube with
superheated steam in countercurrent flow. Following separation of the oil and extraction of the water with benzene, a
total of 29.3 gm, or 8.4%, of the solids was obtained. The
solid pyrolysis residue was found to weigh 228 gm. This
indicated that 91 gm or 26% of the original black liquor
solids were converted to gases. The gases were collected in
gas sample bottles and analyzed by gas chromatograph using
a 7 ft x 1/4 in   stainless steel column packed with silica
gel. The amounts of hydrogen sulfide, methyl mercaptan and
other gases are given below:

Gas found	Volume percent
Air	11
CH_4	5
CH_3SH	4
H_2S	15
CO	30
CO_2	30

In order to determine loss in heating values of the pyrolysis solids, two samples of the residue were analyzed for their heats of combustion. The following results were obtained:

	Heat of combustion
Sample A	5262 BTU/lb
Sample B	5057 BTU/lb
Average	5160 BTU/lb

The heat of combustion of the starting black liquor solids was somewhat more than 6000 BTU/lb, so that from 16 to 20% of the heat value of the solids was lost during pyrolysis. The solids were obtained as a soft black char that was readily suspendable in additional weak black liquor to make a suspension of 55-60% solids that could be burned in a conventional recovery furnace.

A second large-scale experiment was run under similar conditions. A total of 2400 gm black liquor containing 1270 gm solids was pyrolyzed. The oil obtained as described above weighed 109.5 gm, or was 8.6% of the starting solids. Gas chromatograph showed that the oil contained 21% guaiacol or 1.8% of the starting solids.

Part of the oil, 100.0 gm, was distilled through a simple Claisen head distillation apparatus at reduced pressure to give:

Fraction	Boiling point (°C) at 15 mm pressure	Weight (gm)
1	Below 65	0.6
2	65-125	37.7
3	Above 125	34.1
4	Residue	22.8

Gas chromatography showed that Fraction 2 contained 52% guaiacol, or 19.6 gm. This is 1.6% of the starting solids, indicating that some was lost during distillation. A dinitrobenzoate derivative of another distillate similar to Fraction 2 had a melting point of 141.0-143.5°C. Authentic guaiacol dinitrobenzoate had a melting point of 143.0-143.5°C. A mixture of the two had a melting point of 141.0-144.0°C. A further purification was accomplished in the following way: The fraction boiling between 65-125°C/15 mm, 37.7 gm, was treated with aqueous 10% sodium hydroxide, 64 gm, or enough NaOH to react with the guaiacol. The solution was extracted twice with ether to remove neutrals and the less acidic phenols. The ether solution, after drying and evaporation, gave a residue of 17.0 gm. Gas chromatography showed this still to contain 22% guaiacol, so separation was far from complete. The caustic extract was acidified and extracted to give 20.0 gm consisting of about 75% guaiacol by weight.

The guaiacol rich material was analyzed using two different Aerograph chromatographic columns: (1) 10 ft x 1/4 in stainless steel Celanese ester No. 9 on gas chromatography support P, and (2) 12 ft x 1/4 in stainless steel silicone SE-30 on Chromosorb W (AW). The average results, calculated on a weight basis were:

Guaiacol	75%	(b.p. 205°C)
m- and p-Cresols	5%	(b.p. 201°C)
o-Cresol	<1%	(b.p. 191°C)
Phenol	4%	(b.p. 181°C)
m- and p-ethyl phenol	<1%	(b.p. 219°C)

Two unidentified compounds, A and B*

    A about 3%

    B about 5%

Total accounted for, about 94%.

In a final pyrolysis, a sample of concentrated black liquor from the Kraft pulping of a mixture of southern hardwoods was pyrolyzed as described above. Following the previously described work-up and chromatographic analysis, yields of 1.5% of 2,6-dimethoxyphenol and 0.9% of guaiacol were obtained. The relative amounts of these two materials were obtained in about the expected ratios based on the known ratios of syringyl to guaiacyl groups in hardwood lignins of about 3 to 2.

IV. DISCUSSION

As can be seen from the results of the various pyrolysis experiments in the tables, guaiacol can be obtained as the major volatile component of a pyrolysis oil from Kraft black liquor solids. Owing to the nature of black liquor solids and their tendency to foam and expand on heating, provision must be made for breaking the foam into small particles and thus allowing the pyrolytic products to escape before they are destroyed by the high temperatures.

In addition to the use of a screw to crush the foamy solids, the most important variables appear to be temperature control and the use of steam to sweep the volatile products from the pyrolysis zone as rapidly as formed. Temperature control was best obtained by the use of a eutectic salt bath. If the temperature is substantially lower than 400°C, the total amount of oil is reduced, but it is richer in guaiacol. As the pyrolytic temperature is increased, the amount of guaiacol in the oil decreases and the guaiacol is increasingly destroyed. When superheated steam was introduced countercurrently into the black liquor solids flow, the

---

* A was later determined to be 4-methylguaiacol, b.p. 221-222°C and B to be 4-ethylguaiacol, b.p. 234-236°C.

highest yields (2.0-2.1%) of guaiacol were obtained. This
supports the theory that it is important to remove the pyrolysis instantaneously. The steam serves to accomplish the
rapid removal of the guaiacol from the pyrolysis site. The
exact mechanism of the production of guaiacol from coniferous
black liquor is not know. It is clear that the guaiacol
results from cleavage of fragments of the lignin molecule.
This was shown by the use of precipitated Kraft lignin as the
source of organic material in a so-called synthetic black
liquor. The guaiacol could not have arisen in these experiments as the result of pulping reactions. Also, guaiacol
could be obtained from a variety of lignin sources such as
from soda and sulfite pulping.

Finally, it was shown that both 2,6-dimenthoxyphenol
and guaiacol can be obtained by pyrolyzing Kraft black liquor
from hardwood pulping. These products were obtained in
approximately the expected ratio, based on the known syringyl
and guaiacyl groupings in the hardwood lignins.

Thus, it has been shown that by a relatively simple
pyrolysis reaction, potential commercially interesting
amounts of guaiacol and 2,6-dimethyloxyphenol can be obtained
from appropriate black liquors. Fuel value of the liquor
solids is decreased by some 15-20%, but some of this can be
recovered by burning the noncondensible gases given off.
From the pyrolysis, an oil in about 10% yields, based on the
starting solids is obtained. This oil is about 50% distillable, and guaiacol is about 50% of the distillable material.

The process need not interfere with conventional
Kraft furnace operation. If only part of the liquor from a
Kraft mill is dried and pyrolyzed, the soft char can be
readily suspended in additional liquor so that the resulting
solids content is that conventionally burned in a recovery
furnace.

TABLE I

Pyrolysis of Black Liquor with Superheated Steam

Run No.	Pyrolysis conditions							Yields			
								Oil		Guaiacol	
	Black liquor solids pyrolyzed (gm)	Addition rate (gm solids/ min)	Temp. of steam from super- heater (°C)	Temp. of efflu- ent steam (°C)	Average pyroly- sis temp. (°C)	Temp. of direct steam (°C)	Direct steam flow rate (ml condensate/ min)	Oil obtained (gm)	Oil obtained based on solids (%)	Guaiacol in oil (%)	Guaiacol based on solids (%)
1	46.5	0.5	440	395	417	-	-	3.5	7.5	10	0.8
2	51.8	0.5	415	360	387	-	-	3.0	5.8	17	1.0
3	51.8	0.5	372	340	356	-	-	2.4	4.6	22	1.0
4	51.1	0.6	440	395	417	-	-	3.5	6.9	9	0.6
5[a]	51.1	1.0	421	385	403	-	-	4.1	8.0	12	1.0
6[b]	51.1	-	440	395	417	-	-	-	-	-	-
7[c]	51.1	0.7	425	390	407	-	-	2.7	5.3	2	0.1
8	50.4	0.7	430	400	415	-	-	2.8	5.6	14	0.8
9	26.7	0.5	425	350	387	367	9	1.3	4.9	12	0.6
10	34	0.5	415	375	395	400	7	2.4	7.0	12	0.6
11	26.7	-	420	370	390	400	3.5	-	-	-	-
12	51.1	0.7	455	435	445	-	-	3.3	6.5	12	0.8
13	48	0.5	445	425	435	-	-	3.5	7.3	9	0.7
14	48	0.5	500	390	445	-	-	3.3	6.9	8	0.6
15[d]	48	0.5	385	325	355	-	-	2.2	4.6	19	0.9
16[e]	52.1	0.6	495	425	460	-	-	3.3	6.4	11	0.7

a. Screw rotation in runs no. 1-5 was 15 rpm; runs no. 6-16, 30 rpm.
b. Vacuum used.
c. $Na_2S \cdot 9H_2O$ added; 51.1 gm was the amount of original black liquor solids.
d. In runs no. 1-8 and no. 13-15, the pyrolyzate was removed through the lower outlet.
e. In runs no. 9-12 and no. 16, the pyrolyzate was removed through the upper outlet.

TABLE II

Pyrolysis of Black Liquor in a Tube Furnace

| | Pyrolysis conditions[a] | | | Yields | | | |
| | | | | Oil | | Guaiacol | |
Run no.	Black liquor solids pyrolyzed (g)	Addition rate (gm solids/ min)	Pyro- lysis temper- ature (°C)	Oil obtained (gm)	Oil obtained based on solids (%)	Guaiacol in oil (%)	Guaiacol based on solids (%)
17	96	0.8	450	5.9	6.2	10	0.6
18	48	0.6	400	2.8	5.8	12	0.7
19	48	0.6	500	3.2	6.7	5	0.3

a. Screw rotation 30 rpm; temperature in liquor inlet area 250°C (hotplate).

240

TABLE III

Salt as Heat-Exchange Medium for Black Liquor Pyrolysis (Single-Screw Device)

Run no. [a]	Black liquor solids pyrolyzed (gm)	Addition rate (gm solids/min)	Distillate outlet used	Pyrolysis temp. (Salt bath) (%)	Temp. direct steam (°C)	Steam flow rate (ml condensate/min)	Oil obtained (gm)	Oil obtained based on solids (%)	Guaiacol in oil (%)	Guaiacol based on solids (%)
21	152.4	0.9	Lower	470	-	-	14.1	9.3	6	0.6
22	209.5	-	Lower	500	-	-	16.5	7.9	6	0.5
23	228	-	Lower	500	-	-	15.5	6.8	6	0.4
24	57	0.5	Lower	520	-	-	-	-	-	-
25	52.6	0.5	Lower	410	-	-	3.6	6.8	14	1.0
26	311	-	Upper	510	-	-	23.0	7.4	14	1.0
27	102	1.4	Upper	450	-	-	7.3	7.2	19	1.4
28	102.8	1.7	Upper	400	-	-	5.5	5.4	26	1.4
29	101	1.7	Upper	425	-	-	6.1	6.0	20	1.2
30	102.8	0.5	Upper	450	410	5	11.0	11.2	19	2.1
31	50.8	0.6	Upper	500	400	5	5.6	11.0	19	2.1
32	51	0.6	Upper	450	420	5	5.7	11.2	19	2.1
33	25.2	0.3	Upper	450	420	10	2.6	10.3	17	1.8
34	50.8	-	Upper	450	420	7	-	-	-	-
35	30.8	0.5	Upper	450	420	3.5	3.2	10.4	19	2.0
36	39.5	0.4	Lower	450	400	5	4.1	10.4	18	1.9
37	50.9	0.5	Lower	500	390	5	5.5	10.8	11	1.2
38[b]	52	0.6	Lower	450	390	5	1.5	2.9	13	0.4
39[c]	55.6	0.5	Lower	450	390	5	1.3	2.3	20	0.5
40[d]	32.3	0.5	Upper	450	450	5	1.6	5.0	6	0.3
41[e]	51.4	0.4	Upper	450	450	5	4.9	9.5	19	1.8

a. Screw rotation in runs no. 21-23 was 30 rpm; in runs no. 24-41, 40 rpm.
b. The feed in run no. 38 was ORZAN SL-50.
c. The feed in run no. 39 was ORZAN SL-50 + Na_2CO_3.
d. The feed in run no. 40 was spent sulfite liquor.
e. The feed in run no. 41 was Kraft black liquor.

TABLE IV

Salt as Heat-Exchange Medium for Black Liquor Pyrolysis (Double-Screw Device)

| Run no.[a] | Pyrolysis conditions | | | | | Yields | | | |
| | Black liquor solids pyrolyzed used (gm) | Liquor inlet used No. | Addition rate, (gm solids/ min) | Pyrolysis Temp. (°C)[b] | Steam flow rate (ml condensate/min) | Oil | | Guaiacol | |
						Oil obtained (gm)	Oil obtained based on solids (%)	Guaiacol in oil (%)	Guaiacol based on solids (%)
42	27.2	2	0.8	450	5	2.8	10.3	13	1.3
43	266	2	1.0	450	5	27.8	10.1	13	1.3
44	121	1	0.5	450	5	11.9	9.8	18	1.8
45	51.4	1	0.4	500	5	4.9	9.5	19	1.8
46	25	1	0.3	450	6	3.0	12.0	18	2.2
47	36	1	0.3	450	6	3.4	9.4	20	1.9
48	71.5	1	-	450	5	4.6	6.4	26	1.7
49	60.7	1	0.4	350	5	3.4	5.6	33	1.9
50	73	1	0.4	365	5	4.6	6.3	32	2.0
51	77	1	0.4	375	5	5.8	7.5	25	1.9
52	77	1	0.4	385	5	6.2	8.0	24	1.9
53	77	1	0.4	395	5	5.9	7.7	25	1.9
54	77	1	0.4	405	5	5.9	7.7	25	1.9
55	77	1	0.4	415	5	6.5	8.5	23	2.0
56	77	1	0.4	425	5	7.0	9.1	21	1.9
57	77	1	0.5	435	5	7.5	9.7	20	1.9
58	77	1	0.5	445	5	7.5	9.7	20	1.9
59	77	1	0.5	455	5	7.6	9.9	18	1.8
60	77	1	0.5	465	5	7.6	9.9	18	1.8
61	77	1	0.5	475	5	7.8	10.0	18	1.8
62	77	1	0.5	485	5	7.8	10.0	17	1.7

a. In runs no. 46 and 47, dried black liquor solids were used as feed.

b. The direct superheated steam used in this series of runs was heated to the same temperature as the salt bath.

TABLE V

Effect of Variables on Guaiacol Yields

| | Pyrolysis conditions | | | | | | | Yields | | |
| | | | | | | | | Oil | Guaiacol | |
Run no.	Solids pyrolyzed (black liquor) (gm)	Liquor inlet used, No.	Addition rate, (gm solids/min)	Pyrolysis temp. (°C)	Steam flow rate (ml condensate/min)	Screw rotation (rpm)	Oil obtained (gm)	Oil obtained based on solids (%)	Guaiacol in oil (%)	Guaiacol based on solids (%)
63	32.4	1	0.25	325	5	10	1.3	4	37	1.5
64[a]	32	1	0.25	350	5	40	0.5	1.6	0	0
65	79.5	2	0.4	415	4	40	7.0	8.8	14	1.2
66	79.5	2	0.4	350	4	40	5.0	6.3	26	1.6
67	79.5	1	0.4	415	2	40	5.8	7.3	22	1.6
68[b]	74	1	0.4	415	2	40	11.7	-	11	1.8

a. The feed in run no. 64 was obtained by mixing the solid pyrolysis residue from run no. 63 (32 gm) with water (60 ml).

b. In run no. 68, black liquor [140 gm (74 gm solids)] was mixed with a distillation residue [(tar) 10 gm] obtained after guaiacol removal.

243

REFERENCES

1. Pulp and Paper, 49 (7), 31 (1975).
2. W. M. Hearon, W. S. MacGregor, and D. W. Goheen, TAPPI
    45, 28A (1962).
3. T. EnKvist, J. Turunen, and T. Ashorn, TAPPI 45, 128
    (1962).
4. D. W. Goheen, Advan. Chem. Ser. 59, 205-225 (1966).

FOREST FIRES AND THE CHEMISTRY OF FOREST FUELS

R. C. Rothermel

Research Physical Scientist
Intermountain Forest and Range Experiment Station
Northern Forest Fire Laboratory
Missoula, Montana

I. INTRODUCTION

In his paper on the wildland-urban interface fire problem, Butler [1] points out that fires in the home must be under control at all times. Because our homes are fabricated, an ultimate goal can be envisioned where all structures and fabrics are either fireproof or designed so that fires can be easily extinguished. Unlike the urban world, fuels in our forests and rangelands are composed of vegetation that grows and dies in a continuing proliferation of woody material that can and does periodically burn. In the United States during the 5-year period 1970 through 1974, there were an average of 123,703 wildland fires per year, involving an average of about 3 million acres per year.

It has been argued that fire and the forest are not
incompatible. As we often find when natural systems are
examined closely, without man's attempts at control, a cyclical pattern perpetuates the natural system. This changing
view of fire is causing reexamination of the nation's wildland fire control policies. From the dilemma of whether or
not to immediately control all wildland fires has arisen the
concept of fire management, wherein rigid policy is replaced
with options for alternative actions by the land managers
most intimately familiar with the situation.

To make proper decisions and develop fire management
plans, the fire managers are requesting systems for apprais-
ing fuels, weather, and fire potential by systematic methods
that are uniform in their structure but adaptable to local
conditions. The best example of such a system is the
National Fire-Danger Rating System [2], which has been
in operation since 1972. This system provides an evaluation
of the seasonal trends and daily fluctuations in potential
fire severity in the form of relative indexes of fire spread
rate, fire intensity, and an integrated burning index. Nine
representative fuel models that describe the characteristic
differences between grass, brush, timber litter, etc., have
been sufficient for this system.

Systems that require better definitions of fuel properties are under development, and for these more advanced
systems, it is necessary to have a better definition of fuel
properties and the thermal characteristics of fuels. Toward
this end the Forest Service fire research program is attempt-
ing to understand the mechanisms of fire spread and the
influence of the fuels and environment upon fire behavior in
wildland fuels. This chapter discusses recent work on the
heat of combustion of forest fuels and a method for relating
this important parameter to the chemical characteristics of
the fuel.

II. THERMAL CHARACTERISTICS OF FUELS

One of the fundamental properties of fuels that is
needed in the study of fire is the energy that can be expected
to be available from the fuel in the combustion process.

Many researchers now include a weighing system in experiments to provide a record of the mass loss rate as fire consumes the fuel. A simple interpretation of the mass loss rate provides the reaction intensity I_R.

$$I_R = -h \frac{dw}{dt} \qquad (1)$$

where h = heat of combustion (cal/gm) and dw/dt = mass loss rate of fuel (gm/m^2 sec). In the current mathematical model [3] being used for evaluating fire potential, attention has been focused on the mass loss rate and its dependence upon fuel-particle size, fuel loading, fuel depth, fuel moisture, and the mineral content of fuels. Considerable effort has been made to predict mass loss rate using these variables. The heat of combustion h used in Eq. 1 was assumed to be the low heat of combustion obtained from an oxygen bomb calorimeter. This assumption fails to distinguish between energy available as combustible volatiles and energy available in the char that would be released primarily by radiation. The ability to make the distinction between the forms of available energy is necessary to correctly model the mechanism of fire spread.

It is also necessary to relate the expected form of the heat of combustion to the characteristics of the fuel itself. In this case, the intention is to relate the heat of combustion of the flammable volatiles to the initial fuel constituents, namely, the proportions of cellulose, lignin, ether extractives, and inorganic material. It is important to be able to include all four variables, because most of the research in the past has concentrated on cellulose. It has been demonstrated by many researchers that the pyrolysis of cellulose can be altered by the presence of inorganic materials that act as catalysts promoting the formation of char at the expense of the flammable volatiles. This is the basis of the multimillion-dollar fire retardant business that utilizes either diammonium phosphate or ammonium sulfate in thickened slurries dropped from aircraft ahead of wildland fires. Philpot [6] demonstrated that the natural minerals contained in plants could also promote the production of char and hence partially explain differences in combustion in natural fuels.

Could other chemical differences in fuel constituents further explain variation in combustion? It has long been observed that the brown rot from punky wood can be easily ignited and burned primarily in the glowing state with much less flame than sound wood. The brown rot is caused by a fungus that consumes the cellulose and leaves the lignin. Glowing combustion has thus been attributed to lignin. Do the thermochemical properties support this conclusion in a carefully controlled experiment?

The combustion of certain living plants, such as the chaparral of the southwest, palmetto-gallberry of the southeast, and the coniferous forests throughout the country, has been a paradox, because these living fuels burn at moisture levels of 100% or more, and dead fuels will not burn well above 20-30% moisture except in carefully constructed laboratory fuel arrays. It is known that certain living fuels contain large proportions of waxes, terpenes, and oils and so can burn much differently than dead fuels that are lower in these constituents. For instance, pyrolysis is not required for combustion of waxes, terpenes, and oils; simple evaporation is sufficient. The implications of this are not fully understood. Energy is required for both evaporation and pyrolysis, but the amount required in each case must be determined. Chin and DeGroot [4] show that evaporation of volatiles begins at a lower temperature but may not be complete before pyrolysis is well underway.

III. EXPERIMENTAL PROGRAM

To improve our ability to model fire behavior and appraise fuels, clarification of the thermal chemistry of fuels was clearly necessary. A cooperative study was established in 1973 with the University of Montana Wood Chemistry Laboratory under the direction of Professor Fred Shafizadeh to study a wide array of very dissimilar fuels. The objective of the study was to determine the energy available from these fuels and show the division between flammable volatiles and

char at a temperature near the onset of combustion. Furthermore, an attempt would be made to correlate these results
with the initial fuel composition.

The Forest Service gathered and processed the fuels
from throughout the west (Table I) with considerable assistance from the late Dr. Eamor Nord of the Pacific Southwest
Station who, in his efforts to develop nonflammable plants,
was able to supply samples of the saltbushes and other plants
from the southwest.

The thermal chemistry research was performed at the
University of Montana, and a complete report is presented by
Susott and others [5]. The experimental methods were
designed to determine the heat of combustion of the total
sample, the heat of combustion of the char, and the fraction
of total fuel mass converted to char. The heat of combustion
of the flammable volatiles could then be determined by the
difference between the heat content of the total sample and
the heat content of the char.

Briefly, the fraction of char formed was determined by
thermogravimetric analysis; the heat of combustion of the
complete fuel sample, the char, and the extractives were
determined by oxygen bomb calorimetry. All values were corrected to 400°C to simulate the condition near the onset of
combustion. Tabular values of the results of the experimental measurements and the initial fuel constituents are
contained in Tables II-IV.

TABLE I.

Samples Used in Study and Reference Code

Code	Fuel species	Part
1-C	Cellulose	Munktell's filter paper
2-E	Excelsior, Populus spp.	Wood
3-LW	Larch wood, Larix occidentalis	Heartwood
4-PW	Douglas fir, Pseudotsuga menziesii	Punky wood
5-PW-F	Douglas fir, Pseudotsuga menziesii	Punky wood, fine (60 mesh)
6-CS	Chamise, Adenostoma fasciculatum	Stems
7-FWSS	Fourwing saltbush, Atriplex canescens	Stems
8-CVSS	Castlevalley saltbush, Atriplex cuneata	Stems
9-CRS	Creeping sage, Salivia sonomensis	Whole plant
10-PP	Ponderosa pine, Pinus ponderosa	Needles
11-AF	Aspen Populus tremuloides	Foliage
12-CF	Chamise, Adenostoma fasciculatum	Foliage
13-FWSF	Fourwing saltbush, Atriplex canescens	Foliage
14-CVSF	Castlevalley saltbush, Atriplex cuneata	Foliage

TABLE II.

Composition of Fuel Samples (Dry Weight Basis)

Fuel type	Fraction cellulose[a] α_{cell}	Fraction lignin α_{lig}	Fraction extractives α_{ext}	Fraction minerals α_{min}	Silica-free ash S_e
1-C	1.00	0.00	0.00	0.00	0.00
2-E	0.778	0.218	0.00	0.004	0.00381
3-LW	0.770	0.223	0.006	0.001	0.00067
4-PW	0.430	0.534	0.033	0.003	0.00239
5-PW-F	0.197	0.766	0.033	0.004	0.00299
6-CS	0.557	0.391	0.022	0.030	0.0299
7-FWSS	0.645	0.329	0.000	0.026	0.0255
8-CVSS	0.590	0.313	0.031	0.066	0.0643
9-CRS	0.628	0.198	0.101	0.073	0.0484
10-PP	0.539	0.345	0.082	0.034	0.0151
11-AF	0.460	0.368	0.112	0.060	0.0539
12-CF	0.498	0.409	0.051	0.042	0.0390
13-FWSF	0.552	0.307	0.009	0.132	0.128
14-CVSF	0.464	0.327	0.023	0.186	0.180

a.   Measured as total sugar content.

TABLE III

Measured and Calculated Thermodynamic Data

Fuel type	Heat of combustion (cal/gm)					
	Whole fuel		Char[a]		Combined volume[b]	
	$\Delta H^{400°}_{comb}$ (fuel		$\Delta H^{400°}_{comb}$ (char)		$\Delta H^{400°}_{comb}$ (vol)	
	Meas.	Calc.	Meas.	Calc.	Meas.	Calc.
1-C	3853	3853	6848	6977	2833	3213
2-E	4341	4276	7053	6977	2839	2622
3-LW	4391	4321	7004	6977	2528	2667
4-PW	4878	5043	6887	6977	2020	2287
5-PWF	5124	5506	6775	6977	1764	1934
6-CS	4552	4609	7305	6977	2200	2125
7-FWSS	4349	4414	7347	6977	2174	2153
8-CVSS	4211	4348	6906	6977	2035	1969
9-CRS	4590	4360	7207	6977	2312	2358
10-PP	4904	4733	7042	6977	2538	2452
11-AF	4759	4795	7336	6977	2426	2262
12-CF	4757	4711	7184	6977	2178	2123
13-FWSF	4036	4622	7031	6977	2144	2166
14-CVSF	3353	3883	5759	6977	1521	1385

    a. Values in table have been corrected to discount the
weight of ash in the residue. Measured values were determined by experimental methods. Calculated values were
determined by the model described in this paper.
    b. Calories per gram of original fuel.

TABLE IV.

Residue and Char Remaining from Fuels Heated to 400°C

| Fuel type | Mass of char and residue ratioed to initial mass of sample | | |
| | Residue | | Char[a] |
	Meas.	Calc.	calc
1-C	0.149	0.092	0.092
2-E	0.217	0.241	0.237
3-LW	0.267	0.226	0.255
4-PW	0.418	0.398	0.395
5-PWF	0.500	0.516	0.512
6-CS	0.352	0.386	0.356
7-FWSS	0.322	0.324	0.324
8-CVSS	0.381	0.407	0.341
9-CRS	0.389	0.360	0.287
10-PP	0.370	0.361	0.327
11-AF	0.378	0.423	0.363
12-CF	0.401	0.413	0.371
13-FWSF	0.401	0.484	0.352
14-CVSF	0.504	0.544	0.358
15-C[b]	0.076	0.092	0.092
16-C[c]	0.050	0.092	0.092

a. The char values discount the mass equivalent of the minerals, α_{min}, from the mass of residue.

b. Cellulose from Munktell's chromatographic cellulose powder [7].

c. Analytical filter pulp [6].

IV. MODEL DEVELOPMENT

From these data and the conservation of energy equation (Eq. 2) given by Susott and others [5], the following model was developed for predicting the heat of combustion of the fuel, the char, and the fraction of char formed.

$$\Delta H_{comb}^{400°} \text{ (volatiles} = \Delta H_{comb}^{400°} \text{ (fuel)} - \Delta H_{comb}^{400°} \text{ (char)} \times \text{fract. char} \tag{2}$$

$$\Delta H_{comb}^{400°} \text{ (fuel} = 3850 \, \alpha_{cell} + 5860 \, \alpha_{lig} + 7720 \alpha_{ext} \tag{3}$$

$$\text{Fract. char} = (0.0917 + 0.5 S_e^{0.462}) \alpha_{cell} + 0.624 \alpha_{lig} + 0.285 \alpha_{ext} \tag{4}$$

$$\Delta H_{comb}^{400°} \text{ (char)} = 6980 \tag{5}$$

All heat values are calories per gram where α_{cell} = fraction of cellulose in fuel sample, α_{lig} = fraction of lignin in fuel sample, α_{ext} = fraction of ether extractives in fuel sample, and S_e = fraction of ash with silica disregarded.

Equation 3 is a summation of the product of the heat of combustion of the three organic constituents times the fraction of each constituent in the fuel sample. The coefficient for cellulose, 3850, was taken directly from the heat of combustion of pure cellulose. The coefficient for lignin, 5860, was derived from the heat of combustion of the two punky wood samples with corrections for the cellulose and the fraction of lignin. The coefficient for extractives, 7720, was obtained by correcting the heat of combustion values for the extractives given by Susott and others [5] to 400°C. (The standard deviation for the mean value was 3.8%.) Although cellulose composes the largest percentage of the fuel material, its heat of combustion, 3850 cal/gm is much less than that of lignin and only half the value of the ether extractives. The fit of the predicted values to the measured is demonstrated by the clustering of the data along the 1:1

correlation line in Figure 1. Two values appear to be low,
13 and 14; these are saltbushes, and the low prediction is
probably due to errors in the assumption that ash content can
be equated to the mineral content of the plant.

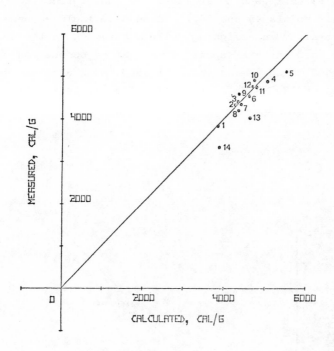

Fig. 1.   Heat of combustion of fuels at 400°C.

Equation 4 is a summation of the fraction of char
formed from each constituent times the fraction of the constituent in the original fuel. If inorganic materials are
present, the fraction of char formed from cellulose is not
constant, as discussed earlier. Using Philpot's [6] criteria
for relating the additional char to the silica-free mineral
content and the data of George and Susott [7] for cellulose
with retardants added, a power function correction factor was
found that accounted for the char produced from cellulose.
The coefficient for lignin was obtained from the punky wood
samples after accounting for the contribution from the

cellulose fraction. The coefficient for extractives was
obtained from TGA data taken by Susott and others [5] on
extractives from six of the fuels. The fit of the char
calculation from Eq. 4 to the measured data is shown in
Figure 2. Supplemental data from George and Susott [7] and
Philpot [6] were used to evaluate the char formed from cellulose (data points 15 and 16). These were included because of
the uncertainty in assigning cellulose values to the sugar
analysis used in this report. The heat of combustion of the
char at 400°C had a mean value of 6980, with a standard
deviation of 396 (5.7% of the mean) for all 14 samples.

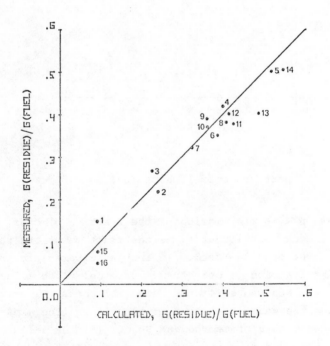

Fig. 2.   Residue from fuels at 400°C.

The heat of combustion of the whole fuel sample shown in Figure 1 has a low measured value of 3353 for Castlevalley saltbush, no. 14, and a high of 5124 for fine, punky wood, no. 5. The majority of points group between 4000 and 5000 cal/gm, which leads one to speculate that the variation in heat of combustion of forest fuels is not large compared to uncertainties in other descriptors. This is misleading when the form of the energy is examined. The residue, which is the char plus the inorganic materials, does vary widely, as shown in Figure 2; from approximately 10% for cellulose to 50% for fine, punky wood, no. 5, and Castlevalley saltbush foliage, no. 14. The high values for these very different materials are caused by the high lignin content in the fine, punky wood (77%) and the high salt content in the saltbush (19%).

The heat of combustion of the flammable volatiles computed from Eq. 2 and compared with the measured values in Figure 3 has measured maximum values less than 3000 cal/gm, which is considerably lower than the 4000 to 5000 cal/gm

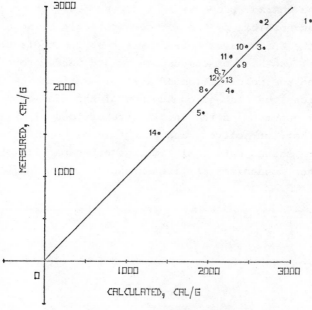

Fig. 3.   Heat of combustion of flammable volatiles at 400°C; calories per gram of initial fuel.

found for the whole fuel sample. The range of data for the
flammable volatiles occupies a larger percentage of the total
range than the whole fuel samples. Very low values are
obtained for the Castlevalley saltbush and the fine, punky
wood. Although cellulose has a low heat of combustion, fuels
with high cellulose content and low mineral content have the
most heat available in the form of flammable volatiles.
These are no. 1 cellulose, no. 2 excelsior, and no. 3 larch
wood. The next groups highest in flammable volatiles are
those high in extractives, no. 9 creeping sage, no. 10 ponderosa pine needles, and no. 11 aspen foliage. These fuels,
assuming they were dry and in a loose, uniform fuel bed,
could produce strong, flaming combustion. This has been
observed for excelsior, larch wood, and ponderosa pine
needles.

V. SUMMARY

    In summary, we have determined that, in order to meet
the needs of fire management agencies, it is necessary to
explore the thermal properties of forest fuels in greater
detail than possible by the methods presently being used. A
study of the energy available from forest fuels directed to
determining the heat of combustion of a diverse group of fuels
has enabled the construction of a simple mathematical model
that predicts the heat of combustion of the flammable volatiles as a function of the initial proportions of cellulose,
lignin, ether extractives, and inorganic materials. This
information should materially improve our capabilities to
describe the mechanisms of fire spread in wildland fuels.

REFERENCES

1. C. P. Butler, The urban and wildland fire interface,
    Fireline April (1976). (Infotech. Publ., San Francisco)

2. J. E. Deeming, J. W. Lancaster, M. A. Fosberg, and
    others. National fire-danger rating system. USDA
    Forest Service Research Paper RM-84. Rocky Mt. Forest
    and Range Experimental Sta., Ft. Collins, Colorado,
    1972.

3. R. C. Rothermel, A mathematical model for predicting
    fire spread in wildland fuels. USDA Forest Service
    Research Paper INT-115. Intermountain Forest and
    Range Experimental Sta., Ogden, Utah, 1972.

4. P. S. Chin and W. DeGroot, Heat release from forest
    fuels. Final Report Cooperative Agreement 12-11-204-1,
    Suppl. 48. Wood Chemistry Lab., Univ. Montana,
    Missoula, Montana, 1975.

5. R. A. Susott, W. F. DeGroot, and F. Shafizadeh, Heat
    content of natural fuels. J. Fire Flammability 6,
    311-325 (1975).

6. C. W. Philpot, Influence of mineral content on the
    pyrolysis of plant materials, Forest Sci. 16, 461-471
    (1970).

7. C. W. George and R. A. Susott, Effects of ammonium
    phosphate and sulfate on the pyrolysis and combustion
    of cellulose. USDA Forest Service Research Paper
    INT-90. Intermountain Forest and Range Experimental
    Sta., Ogden, Utah, 1971.

KETENE FORMATION FROM

THE PYROLYSIS OF CARBOHYDRATES

John C. Kang, Paul H. Chen, and William R. Johnson

Philip Morris Research Center
Richmond, Virginia

I. INTRODUCTION

The formation of ketene from sugars under pyrolytic conditions has long been suspected. Shafizadeh et al [1], in their study of the thermal degradation of 1,6-anhydro-β - D-glucopyranose suggested a dehydration and rehydration rearrangement at C-1, C-2 of the original sugar through a ketene intermediate.

The elimination of the elements of ketene from various cyclic ketone systems upon electron impact is a frequent observation [2, 3]. Since similar structures (e.g., cyclo-pentenones) can occur in the transformation of carbohydrates under pyrolytic conditions [4], it is of interest to study whether ketene is actually formed either as an intermediate or a product.

The presence of ketene in the pyrolyzate of carbohydrates, however, has been difficult to prove. Attempts to
detect ketene by gas chromatographic techniques described in
the literature [5-7] were successful only with pure ketene.
The difficulty presumably was due to the reactions with other
substances in the system, such as water. We therefore have
resorted to indirect methods of analysis based on the known
chemistry of ketene [8]:

$$CH_2CO \; + \; RNH_2 \longrightarrow CH_3CONHR$$
$$CH_2CO \; + \; H_2O \longrightarrow CH_3COOH$$

II. EXPERIMENTAL SECTION

A. Pyrolysis Apparatus

A pyrolysis-gas chromatographic system similar to that
used in the study of amino acids pyrolysis [9] was used with
the following modifications (Figure 1):

1. A larger pyrolysis oven, 9 cm x 14 cm, was used for
maintaining a steady pyrolysis temperature.

2. A side arm with a 1/4 in O.D. was added to the
pyrolysis tube between the connection to the injection port
of the gas chromatograph and the pyrolysis zone. The side
arm was connected to a stainless steel T. One end of the T
was capped with a rubber septum to permit the introduction of
trapping agents. The other end was connected with 1/16 in
Teflon tubing to the carrier gas to insure the complete
transfer of components in the pyrolyzate to the gas chromatography column.

3. A glass wool plug was placed at the intersection of
the side arm and the pyrolysis tube. This plug was saturated
with trapping agents dispensed with a syringe instantly
before the samples were pyrolyzed.

B. Gas Chromatograph

A Varian Aerograph model 1520 equipped with a thermal
conductivity detector and dual columns was used for analysis.
A 22 ft 1/8 in stainless steel column packed with Porapak S,

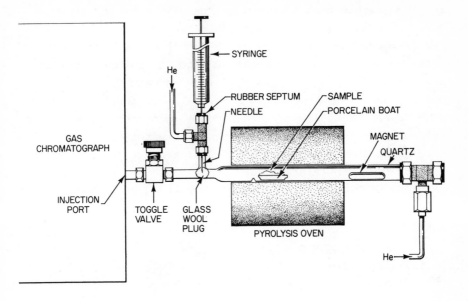

Fig. 1.   Carbohydrate pyrolysis system.

80-100 mesh, was used to separate the pyrolysis products. A
9 ft 1/8 in stainless steel column packed with Chromosorb
101, 80-100 mesh, was used in trapping experiments. The flow
of helium carrier gas was maintained at 80 ml/min. Injection
port temperature was 250°C. Detector cell current was 150 mA.
All connections between the pyrolysis zone and the injection
port except the side arm was maintained at 250°C with a
heating tape. The substances in the pyrolyzate were separated, trapped, and individually identified by a CEC/DuPont
21-104 mass spectrometer.

C. Pyrolysis Experiments

10^{-4} mole of various sugars was pyrolyzed for 90 seconds over a temperature range of 200°-800°C.

D. Trapping Experiments

Both D_2O and n-butylamine were used as trapping agents
for ketene. The former forms deuterioacetic acid, while the
latter forms n-butylacetamide. The ketene yields reported
throughout this work were obtained from the gas

chromatographic peak areas of its n-butylacetamide derivative. The fact that the deuterioacetic acid contains a single deuterium at the methyl group (analyzed by mass spectrometry) suggests that the reaction product is formed by the
attacking of D_2O at the reactive intermediate $CH_2 = C = O$,
rather than by the reaction of D_2O with acetic acid. In the
same manner, mass spectrometric analysis of the reaction
product using a mixture of D_2O and n-butylamine as the trapping agent shows the incorporation of a deuterium in the acetyl group of n-butylacetamide. This again indicates the
reaction of D_2O and/or n-butylamine with an unsaturated
function to form n-butylacetamide (Table I).

TABLE I

Percent Deuterium Incorporation in $CH_2DCONH(D)$-Bu as
Demonstrated in the Pyrolyzates of Sugars at 700°C

	Deuterium incorporation (%)		
	D-Glucose	D-Fructose	D-Galactose
1 deuterium	11.4	16.9	6.2
2 deuterium	3.2	6.5	Not analyzed

A less mechanistically ambiguous experiment involved
the reaction of glucose pyrolyzate with previously distilled
N-deuterated diethylamine. The incorporation of 51-55%
deuterium in the resulting diethylamide $(CH_2DC(=O)NEt_2)$
further verified that the trapping agent was adding across an
unsaturated system (ketene). The inevitable interference of
pyrolytic water in the system precluded 100% incorporation
of the deuterium.

III.   RESULTS AND DISCUSSION

A.   Ketene from the Pyrolysis of Aldohexoses

The production of ketene becomes detectable at 400°C.
Its yield increases with the pyrolysis temperature. The

maximum yeild is at 700°C.   Then the yield declines (Tablè II).
It is known that in the pyrolysis of carbohydrates, concur-
rent pathways lead to the formation of ketones, aldehydes,
and acids [10], which can be further pyrolyzed into ketene.
However, in a study on the pyrolysis of pure acetaldehyde at
450°-525°C, Schuchmann and Laidler [11] reported that ketene
was not found as a final product.   Formation of ketene from
acetic acid under our experimental conditions was checked at
300°-500°C by conversion into n-butylacetamide and was found
to be negligible.

TABLE II
Ketene from the Pyrolysis of Aldohexoses

Temp (°C)	Yield (%)		
	D-Glucose	D-Mannose	D-Galactose
400	1.0	0.6	0.3
500	2.4	1.4	0.5
600	4.2	4.0	2.7
700	16.2	9.2	7.5
800	6.4	3.9	5.4

Bsed on the above considerations, the demonstration of
ketene in the pyrolyzate at temperatures below 500°C indicates that ketene is formed directly from the sugar molecule
rather than from pyrolysis of intermediary acetic acid.   As
it has been shown that conformational and stereochemical
effects which influence the production of anhydro sugars also
influence pyrolytic behavior [12], it is interesting to note
that yields of ketene show an inverse relationship to the
well-known order of anhydro sugar formation:

D-Glucose   <D-Mannose   <D-Galactose

This implies that the transformation of anhydro sugars
adversely affects ketene production.   The degradation of
anhydro sugars under pyrolytic conditions is known to produce
other products [13, 14].

B. Pyrolysis of Anhydro Sugars

Data obtained from the pyrolysis of levoglucosan and
sedoheptulose show that ketene yield from anhydro sugars was
generally lower over all the temperatures than that from the
hexoses (Tables II and III). This difference is especially
prominent at lower temperatures. The overlap observed for
D-galactose is not surprising in that we have noted above
the facility with which it forms anhydro sugars.

TABLE III

Ketene from the Pyrolysis of Anhydro Sugars

Pyrolysis Temp. (°C)	Yield (%)	
	Levoglucosan (1,6-anhydro-D-/ glucopyranose)	Sedoheptulose (2,7-anhydro-D-/ altro-heptulopyranose)
300	0.3	0.4
400	1.5	0.7
500	1.9	1.1
600	3.3	6.1
700	7.4	5.9
800	4.7	2.4

We also observed that in pyrolyses of anhydro sugars at
400°C, more tarry residue was left in the pyrolysis tubes
between the ketene trap and the pyrolysis zone. This observation prompted us to make a comparison of the gas phase
products without ketene trapping.

It is noteworthy that the yields of CO and CO_2 from the
low-temperature pyrolysis of sugars appeared in the order
D-glucose> D-galactose > levoglucosan (Table IV), corresponding to the order of ketene yields from these sugars.

Based on the hypothesis proposed by Shafizadeh [1] that:

$$
\begin{array}{c}
\text{CHO} \\
| \\
\text{HCOH} \\
| \\
\text{R}
\end{array}
\longrightarrow
\left[
\begin{array}{c}
\text{C=O} \\
| \\
\text{CH} \\
| \\
\text{R}
\end{array}
\right]
\longrightarrow
\begin{array}{c}
\text{COOH} \\
| \\
\text{CH}_2 \\
| \\
\text{R}
\end{array}
\longrightarrow
\text{CO} + \text{RCH}_3
$$

CO_2 yield in the pyrolyzate are indicative of the quantity of intermediate formed. Since levoglucosan yields the least quantity of CO_2 and CO as well, sugars that are more readily transformed into levoglucosan would be expected to yield relatively less ketene.

TABLE IV

CO and CO_2 Yields from 400°C Pyrolysis

	Yield (%)		
	D-Glucose	D-Galactose	Levoglucosan
CO	1.76	0.93	0.15
CO_2	1.45	0.72	0.19

C. Ketene Formation from D-Fructose

A comparison of ketene yields from an aldose and a ketose shown in Table V indicates the same yield except at the optimal temperature of 700°C. Thermal degradation of carbohydrates to furan compounds such as 5-hydroxymethyfurfural is well known, and D-glucose was reported to give less yield of this product [15, 16], but more volatile by-products than D-fructose [17]. We found that at 600°C, pyrolysis of D-fructose produces larger quantities of furan compounds that the pyrolysis of D-glucose (Table VI).

To ascertain whether furan compounds were converted to ketene, furan was pyrolyzed under the same conditions and was found to be stable even at 800°C. Therefore, it seems that the formation of furan compounds is competitive with the formation of ketene and ultimately of acetic acid, both of

which are formed from the pyrolysis of glucose and 3-deoxy-
D-erythro-hexosulose [18-20].

TABLE V

Ketene from the Pyrolysis of D-Glucose and D-Fructose

Pyrolysis temp. (°C)	Yield (%)	
	D-Glucose	D-Fructose
400	1.0	1.0
500	2.4	2.1
600	4.2	2.7
700	16.2	8.0
800	6.4	6.2
900	0.2	0.4

TABLE VI

Yields[a] of Furan Compounds from Pyrolysis at 600°C

	D-Fructose	D-Glucose
Methylfuran	42	40
Dimethylfuran	6	2
2-Furfuryl alcohol	128	40

    a. Figures represent concentration of products in
millimeter peak height.

D. Ketene Yield from Certain Structurally Related Hexoses

    To correlate ketene yield with different functions of a
sugar molecule, the pyrolyses of D-sorbitol and D-galacturonic
acid were studied. The results are tabulated in Tables VII
and VIII along with CO and CO_2 yields.

    D-Glucose was shown to yield more ketene than did
D-sorbitol, which, in turn yielded more ketene than did the
uronic acid. These data show that ketene formed by dehydration of an α-hydroxy aldehyde is produced in greater

TABLE VII

Yields of   Products from Pyrolysis at 700°C

Product	Yield (%)		
	D-Glucose	Sorbitol	Galacturonic acid
Ketene	16.2	0.9	0.4
CO	16	18	8
CO_2	3	1.3	10.3

TABLE VIII

Yields[a] of Methanol and Furan from Pyrolysis at 700°C

	D-Glucose	Sorbitol	Galacturonic acid
Methanol	184	448	52
Furan	116	212	70

a. Figures represent concentration of products in
millimeter peak height. Analysis was carried out with a
22 ft x 1/8 in stainless steel column packed with Porapak S,
80-100 mesh.

quantities from a reducing sugar with the preformed aldehyde
group than from a polyol that could degrade to aldehyde
derivatives [18]. Also, the presence of carboxyl groups
could promote alternative degradation pathways, thus reducing
the yield of ketene.

Uronic acids are known to decarboxylate readily [21].
Our experiments indeed show that D-galacturonic acid yields
the greatest quantity of CO_2. This facile loss of C-6 via
decarboxylation causes any ketene formed to be derived from
the remainder of the molecule, where structural features
favorable to its formation are absent.

E. Pyrolysis of Pentoses

Ketene yields from the pyrolysis of certain pentoses
are shown in Table IX. These data show that pentoses that
are more readily dehydrated to furan compounds [22] produce
less ketene. This observation confirms the competition of
the two pathways in the dehydration of the sugars.

TABLE IX

Ketene Yield from the Pyrolysis of Pentoses

Pyrolysis temp. (°C)	Yield (%)			
	D-Xylose	D-Arabinose	D-Ribose	D-Lyxose
400	0.7	0.5	0.7	0.4
500	1.0	1.0	0.8	0.8
600	5.4	5.2	4.0	3.8
700	7.6	7.5	7.5	5.4
800	1.6	1.1	1.6	0.9

F. Ketene Formation from the Pyrolysis of Glycolaldehyde

Having taken into consideration the differences in
ketene yield from all the compounds studied above, the relative importance of the C-5 and C-6 hydroxymethyl group in
conjunction with the ring oxygen becomes apparent.

Glycolaldehyde has been reported to be a product of the
pyrolytic degradation of carbohydrates [23, 24]. In the
present work, we were able to isolate glycolaldehyde from the
pyrolyzates of glucose and galactose, the larger yield being
obtained from the former. Because the glycolaldehyde structure correlates well with the moiety that appears to be
important for ketene formation, its pyrolytic behavior was
studied (Table X). Ketene and water were the only observable
pyrolytic products, and they formed readily, even at 200°C.
Furthermore, the near-quantitative yields obtained suggest
that the mechanism involved is simple dehydration.

TABLE X

Ketene from the Pyrolysis of Glycolaldehyde

Pyrolysis temp. (°C)	Yield (%)
200	70
300	70
500	70

The above experimental data are compatible with a
scheme that describes a side reaction in the pyrolytic transformation of carbohydrates (Scheme I). This may provide a
reasonable explanation for the formation of some products,
e.g., acetic acid. It is likely that this latter substance
is derived from the hydration of ketene (Path B). Though the
present work cannot exclude pathway B-1, it strongly suggests
pathway B-2. However, it is safe to say that carbohydrates

SCHEME

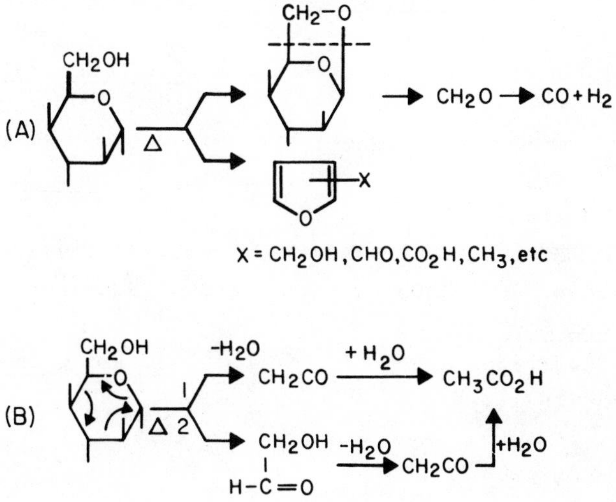

(A)

$$CH_2O \longrightarrow CO + H_2$$

$$X = CH_2OH, CHO, CO_2H, CH_3, etc$$

(B)

$$-H_2O \qquad +H_2O$$
$$CH_2CO \longrightarrow CH_3CO_2H$$

$$CH_2OH$$
$$| \qquad -H_2O$$
$$H-C=O \longrightarrow CH_2CO \quad +H_2O$$

that possess moities that resemble glycolaldehyde tend to give the larger ketene yields. Further mechanistic refinements will obviously require further investigation using other techniques.

REFERENCES

1. F. Shafizadeh and Y. Z. Lai, J. Org. Chem. 37, 278 (1972).

2. A. L. Burlingame, C. Fenselau, W. J. Richter, W. G. Dauben, G. W. Shaffer, and N. D. Vietmyer, J. Am. Chem. Soc. 89, 3346 (1967).

3. R. L. N. Harris and F. Komitsky, and C. Djerassi, J. Am. Chem. Soc. 89, 4765 (1967).

4. R. R. Johnson, E. D. Alford, and G. W. Kinza, J. Agr. Food Chem. 17, 22 (1969).

5. Y. Kikuchi, T. Kikkawa, and R. Kato, J. Gas Chromatog. 5, 261 (1967).

6. J. R. Young, J. Chem. Soc. 2909 (1958).

7. M. J. Dehara, J. Soc. Org. Synth. Chem., Tokyo 20, 730 (1962).

8. R. N. Lacey, Adv. Org. Chem., Methods Results 2, 213 (1960).

9. W. R. Johnson and J. C. Kang. J. Org. Chem. 36, 189 (1971).

10. Y. Houminer and S. Patai, Israel J. Chem. 7, 513 (1969).

11. H. P. Schuchmann and K. J. Laidler, Can. J. Chem. 48, 2315 (1970).

12. G. D. McGinnis, Carbohyd. Res. 31, 183 (1973).

13. J. da S. Carvalho, W. Prins, and C. Schuerch, J. Am. Chem. Soc. 81, 4054 (1959).

14. F. Shafizadeh and G. D. McGinnis, J. Am. Chem. Soc. 16, 273 (1971).

15. R. L. Whistler and J. N. Bemiller in Advances Carbohyd. Chem. 13, 289-329 (1958).

16. R. Weidenhagen, Comp. Rend. Assemblee Comm. Intern. Tech. Sucrerie, London 43-45 (1957).

17. R. R. Johnson, E. D. Alford and G. W. Kinza, J. Agr. Food Chem. 17, (1969).

18. F. Shafizadeh and P. S. Chin, Carbohyd. Res. 46, 149 149 (1976).

19. F. Shafizadeh and Y. Z. Lai, Carbohyd. Res. 42, 39 (1975).

20. F. Shafizadeh and Y. Z. Lai, Carbohyd. Res. 40, 263 (1975).

21. A. S. Perlin, Can. J. Chem. 30, 278 (1952).

22. C. R. Noller, "Chemistry of Organic Compounds" (3rd ed.), p. 434. Saunders, Philadelphia, 1965.

23. A. Qua and I. S. Fagerson, presented in part at the XXIVth Annual Meeting, Institute of Food Technologists, Washington, May, 1964.

24. I. S. Fagerson, J. Agr. Food Chem. 17(4), 747 (1969).

THERMAL DECOMPOSITION AND GAS PHASE ANALYSIS OF

CARBOHYDRATES FOUND IN TOBACCO

H. R. Burton

Department of Agronomy
University of Kentucky
Lexington, Kentucky

I. INTRODUCTION

The development of thermal analysis techniques such as
thermogravimetry (TG) and differential thermal analysis (DTA)
led to studies on the thermal stability of a wide range of
materials. Initially, thermal analysis was used primarily
for investigating the thermal stability of inorganic compounds, but eventually studies were expanded to include
thermally labile organic compounds and naturally occurring
products. These techniques were used to determine the influence of flameproofing chemicals on the thermal decomposition
patterns of cellulose [1, 2] and other carbohydrates [3, 4].
At the same time, dynamic pyrolysis studies of carbohydrates
were performed in order to identify the pyrolysis products
[5, 6] and to identify the major pyrosynthetic pathways of
pyrolysis [7, 8].

Thermal analyses studies have been expanded to more complex systems, such as wood [9-11], tobacco [12-14], and tobacco constituents [15, 16]. A plant material such as tobacco is well adapted for thermal analysis studies, since the major products obtained from it are formed during burning. Therefore, thermal analysis is useful for determining weight changes and the exo- or endothermic reactions that occur during heating. Because tobacco is composed of more than 250 compounds [17] of widely varied structures, the TG curves are not meaningful. Simultaneous gas chromatographic analysis of the evolved gas, however, provides an opportunity to obtain data on the formation of the pyrolysis products. These data aid in identifying and quantifying the products as they are being formed.

From the TG and evolved gas analysis (EGA) data, it is possible to identify the types of reactions that are occurring during thermal decomposition. These simultaneous determinations have been used for identifying the formation profiles of the low molecular weight gases formed during the thermal decomposition of such carbohydrates as cellulose, starch, sucrose, dextrose, and fructose. Also, the influence of the different pyrolysis atmospheres has been investigated to study the influence of atmosphere on thermal decomposition of the carbohydrates. These atmospheres include helium, water-saturated helium, and air.

II. DISCUSSION

Pyrolysis and the thermal decomposition of tobacco and tobacco constituents has been of considerable interest to researchers for the past twenty years. This has been a direct result of determining the contribution of specific tobacco leaf constituents to the formation of health-related smoke components. Many of the earlier studies employed static pyrolysis at 750°C [18], which is the approximate burning temperature of a cigarette. The results can be mis- . leading, since some leaf constituents are distilled from the cigarette at temperatures much lower than the burn temperature of a cigarette [19]. Therefore, it was important to

establish the temperature profile within a cigarette in order
to determine the physical environment that tobacco is subjected to in a burning cigarette. As techniques became
refined, it has been possible to establish well defined
temperature profiles within burning cigarettes[20, 21].
Baker [22] established the temperature profile of a cigarette
for both the solid and the gas phase during a 2 sec duration
puff (Fig. 1). The purpose for presenting the figure is two-
fold. First, it shows the shape of the temperature profile
of the cigarette, but more importantly, it shows the length
of the reaction zone. This length indicates that there could
be sufficient time for slower reactions to occur at lower
temperatures. Additional studies by Baker [23] have shown
that defined areas within the burning cigarette have higher
concentrations of such constituents as CO, CO_2, and methane.
Figure 2 contains the temperature, oxygen, carbon monoxide,
and carbon dioxide profiles obtained in a burning cigarette
[23]. It should be noted that the oxygen content behind the
burning cone is depleted, which shows that reactions are
occurring in an inert atmosphere. Carbon monoxide has two
areas for maximum formation, and carbon dioxide has maximum
formation at a relatively low temperature. The profiles for
methane and ethane are similar to those for both CO and CO_2.
Since areas of high concentrations of these gases exist
within a burning cigarette, there was need to determine if
these levels of maximum formation occur during the thermal
decomposition process.

Thermal analysis of tobacco seemed to be a good system
for studying the thermal reactions that are occurring during
its thermal degradation. The DTG and EGA traces in Figure 3
show the complexity of thermal decomposition of tobacco,
since there are six distinct dw/dt maxima exhibited by DTG
and these maxima correspond closely to those obtained from
EGA in a helium atmosphere. As expected, different types and
varieties of tobaccos result in quantitative differences in
both the DTG and EGA curves [12, 13, 24].

From the thermal analysis data of this complex material several questions were raised. What thermally degraded
leaf materials were responsible for the characteristic DTG

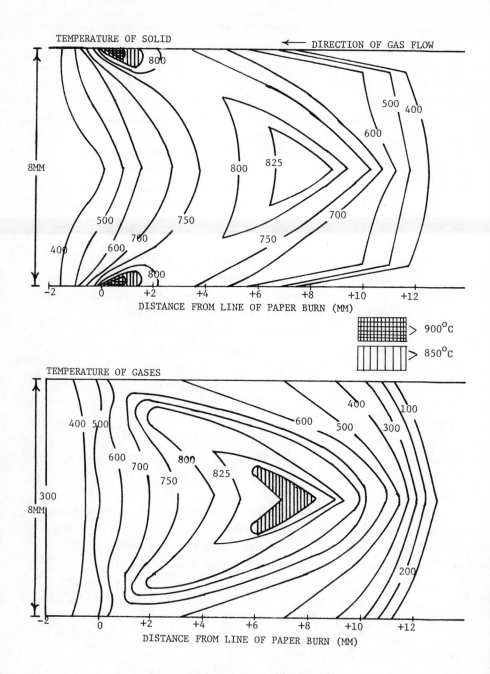

Fig. 1.   Temperature (°C) distribution in the coal,
2.0 sec. after start of a puff [22].

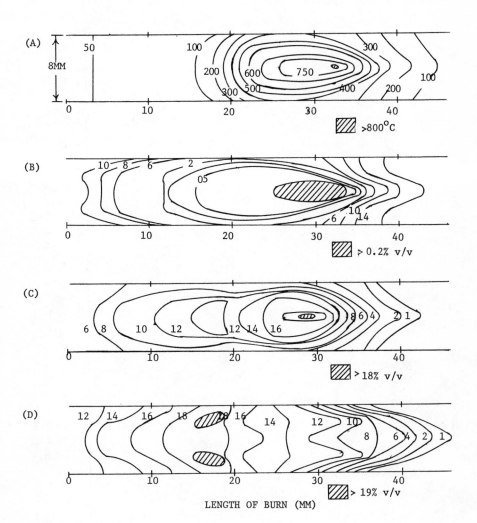

Fig. 2.   Temperature (°C) and gas concentration
(% v/v) contours.   (A) temperature (B) oxygen (C) carbon
monoxide (D) carbon dioxide [23].

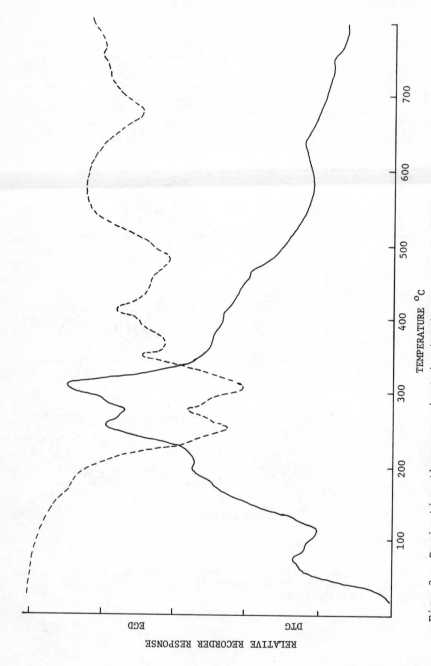

Fig. 3. Derivative thermogravimetric (DTG) curve and effluent gas detection (EGD) curve for 1R1 tobacco. DTG ———, EGD – – – .

and EGA patterns? What types of interactions between leaf
constituents were involved during thermal degradation? What
types of pyrolysis products were responsible for the weight
loss and evolution of gases? Of the three, the last will be
considered in greatest detail.

Standard thermal analysis texts briefly discuss
evolved gas and analysis of the individual gases [25, 26],
and many thermal analysis instruments have been designed to
allow for the analysis of individual gas phase components.
However, the literature contains few references for the stim-
ultaneous TG and the analysis of complex effluent gas mixtures [27, 28]. Because of its availability, gas chromatography was used in our laboratory for the separation and
detection of individual pyrolysis components. A differential
thermal analysis unit was conveniently coupled to a gas
chromatograph via a 5.0 ml gas-sampling loop. By using a
Carbosieve B column and a thermal conductivity detector, it
was possible to analyze the evolved gas for CO, CO_2, and H_2O.
Employing a 6 ft Poropak Q column and a flame ionization
detector, it was possible to analyze for hydrocarbons such
as methane, ethene, ethane, propane, propene, butane, butenes,
isoprene, benzene, and toluene and the low molecular weight
carbonyl compounds such as formaldehyde, acetaldehyde, acrolein, acetone, and butanone.

The system, because of its design, does not allow for
the continuous monitoring of the gas phase constituents as
they were being formed. Therefore, 5.0 ml samples were taken
at 25°C intervals between 50° and 900°C. Because of the time
required for separation by gas-liquid chromatography, a minimum of fourteen separate tobacco samples were required. A
plot of recorder response versus temperature for each of the
components analyzed resulted in the formation profile. In
constructing the formation profiles, it is possible to determine the temperatures for maximum formation and the relative
concentration of the products.

Formation profiles for CO and CO_2 from the thermal
decomposition of tobacco are shown in Figure 4 [29]. The CO
maxima at 250°, 310°, and 700°C correspond to both the DTG
and EGA maxima. The maxima also correspond quite well to

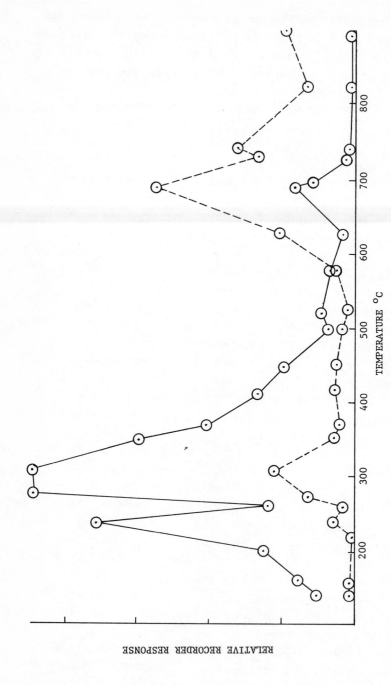

Fig. 4. Formation profiles for carbon dioxide and carbon monoxide from the analysis of the effluent gas from DTA of 1R1 tobacco. CO – – – , CO$_2$ ———.

those maxima found in a burning cigarette (Fig. 2). Except
for quantitative differences, the profile for CO_2 exhibits
similar maxima. Profiles from a similar study agree qualitatively with these profiles, even though the heating rate
employed by Baker was seven times greater [30].

Profiles obtained for other low molecular weight
hydrocarbons also have been obtained from the thermal degradation of tobacco. The data show that the temperature for
maximum hydrocarbon formation occurs between 475-525°C. The
aldehydes are formed at lower temperatures than the hydrocarbons (300°C), and the other volatile components, such as
acetone, butanone, benzene, and toluene, have multiple
maxima between 350° and 550°C [31].

From the above studies, the maxima for the profiles
corresponded from DTG and EGA maxima. The question was
posed as to what plant components were responsible for the
resulting profiles. It therefore seemed appropriate to study
the thermal decomposition patterns of individual tobacco
components, to determine if they could contribute to the
formation profiles obtained from tobacco. The obvious plant
constituents to investigate are the carbohydrates, since
they comprise 55% dry weight of tobacco [33]. The studies
involved the polysaccharides (cellulose and starch) a
disaccharide (sucrose), and the monosaccharides (dextrose
and fructose plus maltose, lactose, pectin, and cellobiose).

The pyrolysis of carbohydrates has been one of the
most extensively studied of any class of organic compounds.
Literature references for the pyrolysis of carbohydrates and
analysis of their products are beyond the scope of this
review; however, research prior to 1968 has been covered in
an excellent review by Shafizadeh [33]. Originally, the
number of compounds analyzed were limited, but the introduction of capillary column chromatography and gas chromatography-
mass spectroscopy dramatically expanded the number of
products that can be simultaneously analyzed [5]. Also,
studies using labeled carbohydrates have increased the understanding of the mechanisms involved during the pyrolysis of
carbohydrates [34, 35].

Almost all of the pryolysis studies up to this time
have used isothermal pyrolysis conditions. Heyns [36] has
used several pyrolysis temperatures, but these are not the
same as a dynamic pyrolysis system. The studies of isothermal
pyrolysis are useful for determining the products that may be
formed during the thermal degradation of carbohydrates. However, they are of little value for predicting the reactions
that are occurring within carbohydrate matrix as it is being
heated in a dynamic environment. There is meager information
in the literature on the profiles of pyrolysis products
formed during the dynamic thermal degradation of carbohydrates. The profiles that have been reported are CO, CO_2,
and water and therefore do not include the other twenty major
pyrolysis products [37, 38]. There is only one recent
report on the thermal decomposition of cellulose, but it
included only the profiles for CO and CO_2 formation [39].

In order to establish the formation profiles for the
low molecular gases, studies were initiated on the thermal
decomposition of the carbohydrates. Figure 5 contains the TG
curves for five of the carbohydrates used for this study.
They are presented in order to define the weight change patterns and will be referred to during the discussion of the
formation profiles.

Except for cellulose, the formation profiles of carbon monoxide obtained in helium from cellulose, starch, dextrose, and fructose do not show any large maxima that can be
associated with the weight loss of the carbohydrates (Fig.6).
The temperature for maxima formation of CO from cellulose is
65°C lower than that reported by Baker [39]; however, the
heating rate used by the latter is twenty times greater than
the heating rate of 6°C/min used in this study. One would
expect the maxima to shift to higher temperatures at faster
heating rates. There is an increase in CO formation at
300°C, which shows CO is being formed, as expected, during
the major weight change region of the thermogram. There is .
no large decrease in CO formation above 400°C, which also
indicates at elevated temperatures the carbohydrate residue
contributes significantly to its formation.

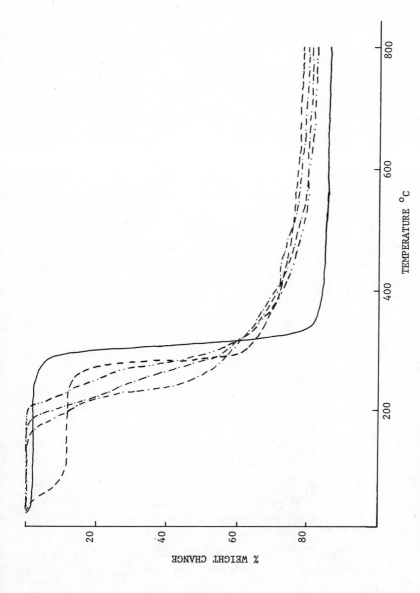

Fig. 5. Thermogravimetric curves of carbohydrates. Cellulose (———), starch (-- -- --), dextrose (- • - • -) fructose (- - - -), sucrose (- -- • • -- -).

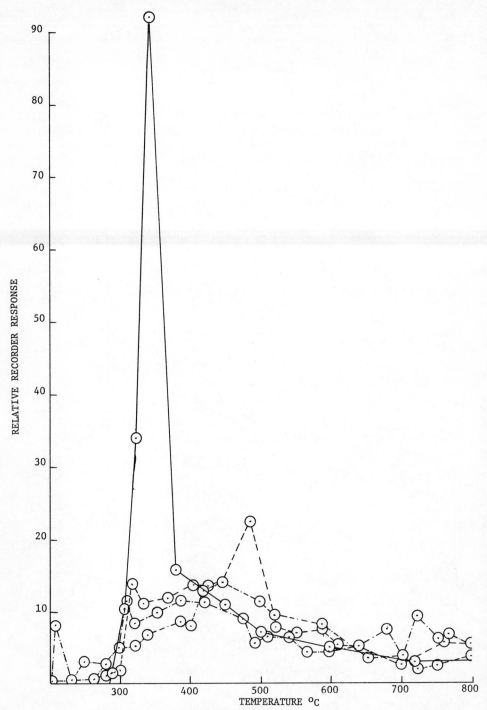

Fig. 6. Formation profiles for CO from carbohydrates.
Cellulose (————), starch (- - - -), dextrose (- · - · -),
fructose (- - · - -).

Carbon dioxide has profiles (Fig. 7) that are similar to those of carbon monoxide; however, the maxima between 300°C and 400°C are more pronounced for the CO_2 profiles. Above 400°C, the level of CO_2 drops rapidly, even though the level is somewhat higher than that of CO. This is in agreement with the data obtained from other thermal decomposition studies [37, 38]. From these data, one can conclude that the contribution of carbohydrate residue is not significant at temperatures above 500°C when it is compared with the total amount of CO_2 formed during thermal degradation. Even though the profiles for CO and CO_2 show that the carbohydrates contribute to the CO and CO_2 obtained from tobacco, it is impossible to extrapolate their contribution to the level of CO and CO_2 formed from tobacco. This can only be accomplished by additional studies on the interaction between the carbohydrates and other plants materials during their thermal degradation.

The methane profiles (Fig. 8) from carbohydrates have formation maxima between 475° and 550°C. Methane is formed between 300° and 400°C, but it is quite low in comparison the level of formation at higher temperatures. Even though there is only a small weight change occurring between 400° and 600°C, methane formation is a result of this weight loss. The levels of methane obtained from the polysaccharides are lower than those for mono- and disaccharides. This may be the result of lower residue weights for the polysaccharides in comparison to the mono- and disaccharides (Fig. 5).

Profiles for the other hydrocarbons are similar to that for methane, since their maximum formation is above 400°C; however, their maxima are 100°C lower than that for methane. Rather than present the profiles for all of the hydrocarbons, only the levels and temperatures of maximum formation will be reported. Data reported in Table I for seven of the low molecular weight hydrocarbons are presented to show only the differences in the temperatures for maximum formation and relative levels of formation. The table contains data from the nine carbohydrates that have been studied. It should be noted that cellobiose and lactose are not found in tobacco, but they were included in the study to

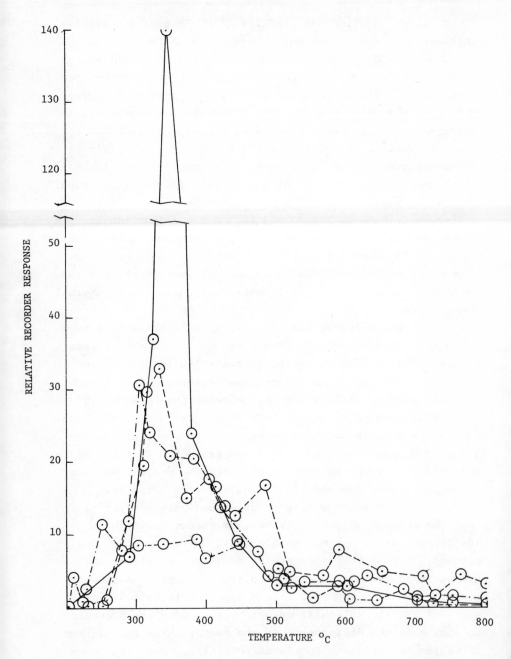

Fig. 7. Formation profiles for CO_2 from carbohydrates; cellulose (———), starch (- - - -), dextrose (- · - · -), fructose (- - · - -).

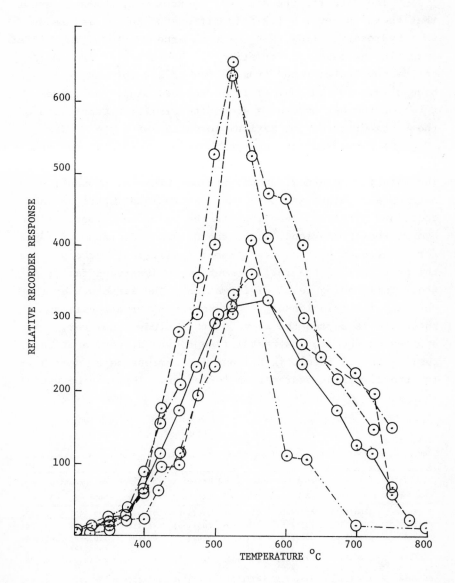

Fig. 8. Formation profiles for methane from carbohydrates: cellulose (——), starch (- - - -), dextrose
(- · - · -) fructose (- - · - -), sucrose (- - ·· - -).

determine if minor differences in the structure of disaccharides would yield different formation profiles.

The data for the simplest hydrocarbon, methane, shows
that there is over a 3.2-fold difference in the levels of
this hydrocarbon from those of the carbohydrates. The differences in the level of formation for hydrocarbons from different carbohydrates range from as low as 2.8 for ethene to a
high of 8.3 for butene. This is unique, since the structures of the carbohydrates are quite similar. Comparison of
their temperatures for maximum formation with the TG data
(Fig. 5) shows that the hydrocarbons are formed at the end of
the major weight change of thermal decomposition. Therefore,
the data for methane, ethene, ethane, propene, propane,
butene, and butane indicate that the residue (char) and not
secondary pyrolysis reactions is the major contributor of
hydrocarbon formation. Since there is a wide range of levels
for individual hydrocarbons from the different carbohydrates,
the structure of the residue must be an important factor in
their influencing level of formation. The structure of the
char and its relationship to hydrocarbon formation is not
known and is beyond the scope of this study. However,
studies on the factors that influence the structure of the
residue should result in a means for increasing or decreasing the levels of hydrocarbon formation. This may be useful

TABLE I.

Levels of Hydrocarbon Formation from Thermal Decomposition of Carbohydrates

| Carbohydrate | Hydrocarbon formation level [temperature of maximum formation (°C)] | | | | | | |
	Methane	Ethene	Ethane	Propene	Propane	Butenes	Butane
Cellulose	3200(475)	460(460)	410(450)	240(460)	100(440)	70(440)	42(440)
Starch	2050(500)	233(475)	250(450)	205(450)	96(450)	70(450)	49(450)
Pectin	2700(500)	300(475)	450(475)	250(475)	440(475)	270(250) 200(475)	160(475)
Lactose	3000(550)	275(425, 500)	540(500)	310(425)	141(425)	200(425)	88(425)
Sucrose	3500(550)	345(475)	470(475)	188(475)	150(475)	36(475)	72(450)
Maltose	1760(475)	260(450)	430(450)	184(450)	110(450)	58(450)	58(425)
Cellobiose	2550(525)	225(475)	365(475)	255(450)	126(475)	86(425)	-
Dextrose	4448(525)	730(525)	890(525)	435(475)	265(475)	112(450)	120(475)
Fructose	1370(550)	390(475)	135(450)	84(450)	58(450)	24(425)	28(425)

for the production of synthetic gases from carbohydrate wastes.

The formation profiles for the low molecular weight adldehydes show a wider range of levels of formation and a wider range of maximum formation temperatures than for the hydrocarbons. This is shown by the profiles for formaldehyde from five different carbohydrates presented in Figure 9. Of the five carbohydrates, fructose is the best precursor for formaldehyde and has a temperature of maximum formation at 225°C. The other temperatures for maximum formation are below 400°C, which shows that formaldehyde formation is occurring during the major weight loss transition. One cannot postulate the mechanism as being a result of initial thermal decomposition arising from the solid or liquid phase or as a result of secondary pyrolysis. However, one would predict that formaldehyde is formed from the residue, since both cellulose and starch, which have rapid weight loss, also have the lowest levels of formaldehyde formation. The TGA for the mono- and disaccharides show that the weight loss occurs at lower temperatures, and it has been postulated that these weights losses are a result of dehydration and polymerization [33]. This supports the postulate that formaldehyde formation is favored when carbohydrates decompose via dehydration and polymerization.

Data in Table II include the maximum levels of formaldehyde for cellulose, starch, sucrose, fructose, and dextrose, plus four additional carbohydrates. The data show that there is over a 200-fold difference in the level of formaldehyde that can be found from different carbohydrates. Again, the wide range of formaldehyde levels is a result of the different decomposition patterns of structurally similar carbohydrates.

Maximum formation for acetaldehyde does not have the wide range of levels as formaldehyde, yet there is an eightfold difference in its levels of formation. Cellulose being the best precursor and fructose the poorest precursor for acetaldehyde is the opposite of what was observed for formaldehyde. This indicates that the pyrolytic pathways for these two aldehydes are quite different. One of the proposed

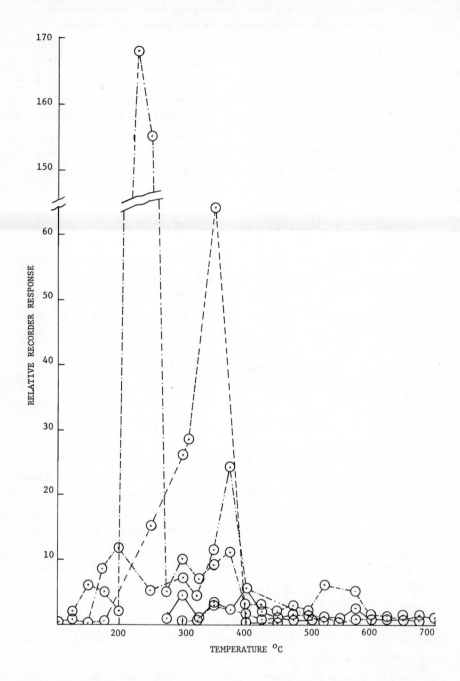

Fig. 9. Formation profiles for formaldehyde from carbohydrates: cellulose (———), starch (- - - -), dextrose (- · - · -), fructose (- - · - -), sucrose (- - ·· - -).

TABLE II.
Levels Aldehyde and Ketone Formation from Thermal Decomposition of Carbohydrates

| Carbohydrate | Aldehyde-ketone formation level [temperature of maximum formation (°C)] | | | | |
	Formaldehyde	Acetaldehyde	Acrolein	Acetone	2-Butanone
Cellulose	65(325)	800(350)	2032(350)	1450(340)	740(340)
Starch	8(327)	145(325)	432(400)	136(175)	346(300)
Pectin	2050(380)	435(250)	265(380)	335(400)	270(400)
Lactose	151(400)	260(425)	275(400)	340(400)	380(400)
Sucrose	920(200)	140(375)	148(325)	270(425)	250(375)
Maltose	270(350)	245(325)	400(375)	-	245(325)
Cellobiose	285(250)	285(350)	265(475)	576(350)	345(350)
Dextrose	238(225)	255(375)	800(400)	830(400)	875(400)
Fructose	1680(225)	96(325)	480(425)	176(325)	188(325)
				163(425)	127(425)

mechanisms for acetaldehyde involves cracking of 1-hydroxy-2-propanone into acetaldehyde and formaldehyde [40]. The data from cellulose and fructose indicate that this pyrolytic route is not important for these carbohydrates, since the levels of these two aldehydes do not parallel each other. Another route, such as decarbonylation of pyruvaldehyde, would seem to be more operative. It is apparent additional studies are required to clarify the mechanism for acetaldehyde formation.

Table II also contains data for acrolein formation. Not only is acrolein a major pyrolysis product, it also has ciliostatic activity. Except for cellobiose, the formation profiles for acrolein are in the same temperature ranges as those for formaldehyde and acetaldehyde. Again, there is a large difference between the high and low levels of acrolein formation. It is of interest to note that cellulose is the best precursor for acrolein, and cellulose is the most predominant carbohydrate in tobacco. The fragmentation pattern responsible for acrolein formation, postulated by Byrne et al. [41], involves the scission of the hexose into pyruvaldehyde and acrolein. Unfortunately, pyruvaldehyde was not analyzed in this study. It would be interesting to determine if the levels of these two aldehydes are consistent with the postulated mechanism. Again, this data could lead to

additional confusion, since pyruvaldehyde can undergo secondary decomposition.

The relative levels of acetone and butanone presented
in Table II again show the wide range of formation levels
that can be obtained from different carbohydrates. Cellulose is one of the better precursors for acetone and butanone, whereas, starch and fructose are the poorest precursors. It should be noted that the profiles for both acetone
and 2-butanone from fructose have double maxima. This is not
unique, but this is the first time it has been observed in
this study. These double maxima add to the complexity of the
decomposition pathways that may occur during the thermal
degradation of carbohydrates.

From the data obtained for the aldehydes and ketones,
a generalized statement can be made concerning their formation. Formaldehyde is formed in largest quantities during
the carbohydrates' initial weight loss. This seems to indicate that it is being formed from the substrate rather than
being a secondary pyrolysis product. The other four aldehydes and ketones are formed during maximum weight loss or
when the weight loss is quite low. Therefore, at higher
temperatures, these compounds most likely are being formed
directly from the thermally degraded residue.

Now that the formation profiles for the low molecular
weight gas phase constituents have been established, one must
pursue the study further in order to determine how the different atmospheres influence the formation of these products.
the obvious was to investigate the effects of oxygen on their
formation, since in a burning process, oxygen must influence
the level of compounds found in the gas phase. Zero grade
air was used in order to minimize the influence of impurities
that may be present in compressed air. It must be noted that
the flow of air (30 ml/min) is in greater excess than would
be found in a smoldering cigarette. Even so, an excess of
oxygen should provide information of its influence on the
formation of gas phase constituents.

A typical example of the influence of oxygen on hydro-
caron formation is shown by the methane profile from cellulose (Fig. 10). Since the profiles for methane from the

other carbohydrates are similar, only the profile from
cellulose will be presented. There are qualitative and
quantitative differences in these two profiles. In air, the
temperature for maximum formation is at 315°C, which is over
20°C lower than that in helium. The maximum in air is higher
than in helium; however, the total quantity of methane
evolved is much greater in helium than in air. In air, the
increased methane formation at lower temperatures indicates
that the exothermic oxidization reactions [42] that are
occurring enhance the methane formation. One would expect
that methane formation would be quite low in the presence of

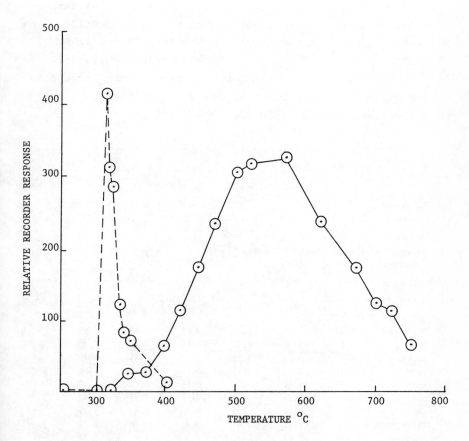

Fig. 10. Formation profiles for methane from cellulose; He (———), air (- - - -).

oxygen. However, at 300°C, carbohydrates do not undergo combustion and therefore the oxygen may only act as a catalyst for the radical reactions responsible for methane formation. This is supported by TG data, since they show that air in comparison with helium has no influence on the weight change between 200° and 400°C. At higher temperatures, there is no methane formation, which indicates that there is sufficient oxygen to completely oxidize the carbon residue to CO, CO_2, and H_2O.

The ratios of methane formed in air versus helium from the carbohydrates are in Table III along with the ratios of the six other low molecular weight hydrocarbons. Air, as expected, reduces methane formation from six of the carbohydrates; however, three carbohydrates are the exception. The temperatures of maximum formation are from 25° to 125°C lower in air than in helium. As seen by the ratios for methane, there is no trend that will explain the large differences in methane formation. Again, it is possible that methane formation is influenced by the structure of the residue.

Ratios for the other low molecular weight hydrocarbons in Table III show wide differences in their levels of formation. Air can effect a 300% increase or a 94% decrease in hydrocarbon formation. Cellulose is the only carbohydrate

TABLE III.
Ratio of Levels of Hydrocarbon Formation from Thermal Decomposition of Carbohydrates in Air versus Helium

Carbohydrate	Formation level ratio[a]						
	Methane	Ethene	Ethane	Propene	Propane	Butene	Butane
Cellulose	1.31	2.24	1.35	2.57	1.84	3.66	–
Starch	0.50	0.34	0.34	0.26	0.44	0.39	0.29
Pectin	0.89	0.19	0.77	0.20	0.41	1.30	0.38
Lactose	0.10	0.33	0.11	0.10	0.09	0.04	0.05
Sucrose	0.56	0.26	0.63	0.16	0.47	0.14	0.08
Maltose	0.08	0.17	0.19	0.41	0.33	0.14	0.34
Cellobiose	1.31	0.28	0.34	0.27	0.23	0.20	–
Dextrose	0.06	0.11	0.07	0.08	0.06	0.15	0.09
Fructose	1.82	0.24	2.37	0.30	1.28	0.46	0.11

a. Detector response in air/detector response in He.

n which air effects an increase for all of the hydrocarbons. For cellulose, it is also of interest to notice the air increases the ratios of the unsaturated hydrocarbons more than ratios of the saturated hydrocarbons. Air decreases the levels of all the hydrocarbons from starch, dextrose, fructose, lactose, sucrose, and maltose. Fructose is rather unique, since air effected an increase in the levels of three of the saturated hydrocarbons while reducing the levels of the four other hydrocarbons.

The data for the aldehydes and ketones also show some interesting trends. A typical profile, acetaldehyde from cellulose, is presented in Figure 11 and shows that air dramatically enhances acetaldehyde formation and does not shift the temperature of maximum formation. For this profile, there is over a fourfold increase in acetaldehyde formation. At this time, the role of oxygen in the enhanced formation of acetaldehyde is not known; however, one can postulate several modes of action. The increased acetaldehyde may be the direct result of oxidation of unsaturated hydrocarbons, saturated hydrocarbons, or alkyl radicals in the gas phase. Also, the oxidation may be occurring at the surface of the solids substrate, which would alter the decomposition mechanisms. An alternative mechanism to the increased acetaldehyde formation may be due to the elevated temperatures due to the exothermic reactions that are occurring. Data that are available at this time do not support or eliminate any of the above mechanisms. Additional research in this area is certainly needed to clarify the enhanced formation of acetaldehyde in the presence of oxygen.

The ratios for acetaldehyde from the nine carbohydrates investigated are presented in Table IV. Air increases the level of this aldehyde from six of these carbohydrates. It should be noted that five of these carbohydrates are found in tobacco [42] and that the profiles from tobacco are similar to acetaldehyde profiles from the carbohydrates [43]. The ratios for the four other aldehydes and ketones do not show the number of increases that were obtained for acetaldehyde (Table IV). Data for formaldehyde show that air enhances its formation from the three polysaccharides and

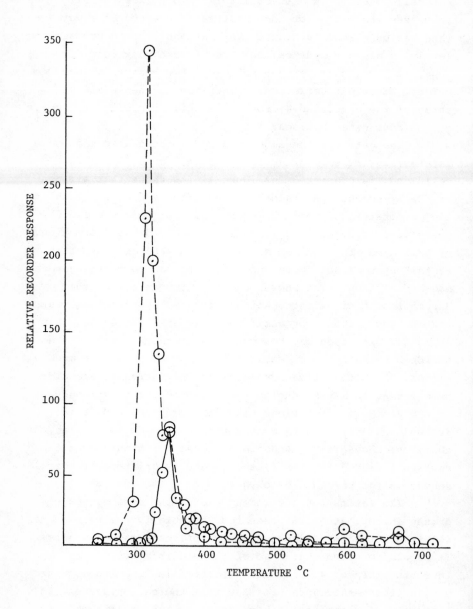

Fig. 11.   Formation profiles for acetaldehyde from cellulose; He (———), air (- - - -).

TABLE IV.

Ratio of Aldehydes and Ketone Formation from Thermal Decomposition of
Carbohydrates in Air versus Helium

Carbohydrate	Formation level ratio[a]				
	Formaldehyde	Acetaldehyde	Acrolein	Acetone	Butanone
Cellulose	5.38	4.30	–	4.69	2.50
Starch	1.75	2.21	1.16	0.76	1.24
Pectin	1.68	0.40	0.62	0.66	0.56
Lactose	0.08	0.81	0.95	0.61	0.33
Sucrose	0.01	1.36	3.92	0.38	0.48
Maltose	0.19	0.05	0.44	–	0.02
Cellobiose	0.25	1.68	0.41	0.61	0.38
Dextrose	0.41	1.41	–	0.46	0.26
Fructose	0.02	1.77	–	1.28	0.88

a. Detector response in air/detector response in He.

decreases its level of formation from the mono- and disaccharides. This is the only case where there is a difference
in profiles between mono or disaccharides and polysaccharides. The ratios for the two ketones show that the use of
an oxidizing atmosphere reduces the formation of these two
compounds from the majority of the carbohydrates. However,
except in the case of maltose, the levels of ketones are not
dramatically reduced.

The influence of an oxidizing atmosphere on the
formation of aldehydes and ketones from two of the three
polysaccharides shows that air generally enhances the formation of aldehydes and ketones from those carbohydrates that
undergo rapid volatilization. By contrast, such carbohydrates as maltose and dextrose, which undergo slower weight
changes, have lower ratios of aldehydes and ketones when
they are thermally degraded in the presence of air.

The last area of study to be presented is the influence of water vapor on the formation of the low molecules
weight gases from carbohydrates during their thermal decomposition. Inclusion of the influence of water is important,
since in the burning process, the quantity of water present
in the atmosphere varies with the relative humidity. The
amount of water in the atmosphere, therefore, may be an

important factor for influencing the composition of the
pyrolysis products. To study the influence of water on the
level of formation of the low molecular weight products,
helium was saturated with water vapor by passing it over
water at 20°C and gave a concentration of approximately 2%
v/v. An inert carrier gas was used in order to study only
the effect of water and not the combined influence of an oxi-
dant and water, as would be the case if water-saturated air
were used.

The profile for carbon monoxide (Fig. 12) is typical
of the profiles that were obtained from the other carbohydrates. The profile for CO in helium is included for comparison with the profile of CO obtained in the helium-water atmosphere. The figure does not include the portion of the
profile from ambient temperature to 200°C, since the level of
formation is minimal at these temperatures. At the mid-
range temperatures, water vapor has little influence on the
formation of carbon monoxide. There are small differences,
but they are significant in comparison to the differences
obtained above 800°C. In water-saturated helium, the level
of CO is increased dramatically between 800°C and 950°C. The
large increase in CO formation is apparently due to the
reaction of water vapor with the carbon residue. The hydrogen levels, which have not been presented, are also increased
at these elevated temperatures. This indicates that the
water is the oxygen source for the carbon monoxide. The TG
data for the carbohydrates support the postulate that water
is reacting with the carbon residue, since a 3% weight change
occurs between 850° and 950°C [44]. The CO formation profiles from the starch, cellulose, and dextrose thermally
degraded in $He-H_2O$ all have large maxima at approximately
900-950°C.

Above 800°C the carbon monoxide level for tobacco is
also increased when it is thermally decomposed in the
presence of water [44]. This indicates that the reaction of
water with carbon residue at elevated temperatures is a general reaction that involves a common mechanism. The level of
carbon monoxide formed is dependent upon the quantity of

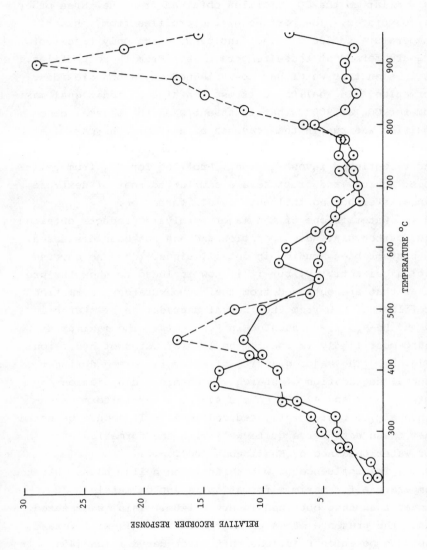

Fig. 12. Formation profiles for carbon monoxide from fructose. He(——), He-H$_2$-(- - -).

residue available during thermal decomposition at these
temperatures.

The formation profile for CO_2 (Fig. 13) from dextrose
is similar to the CO_2 profiles obtained from the three other
carbohydrates. The portion of the profiles from ambient
temperature to 200°C is not included, since only traces of
CO_2 are formed at these temperatures. From these profiles,
it is seen that water has no influence on carbon dioxide
formation. In helium, dextrose contains an additional maximum for CO_2 at 900°C that is absent when the thermal decomposition was run in the presence of He-H_2O. The maxima at
900°C is characteristic for only dextrose and presently is
under continued investigation. Profiles for CO_2 from cellulose, starch, and fructose are similar to that of dextrose,
since water has no influence on CO_2 formation.

Water is one of the major pyrolysis products obtained
from carbohydrate decomposition and has not been discussed.
It has not been studied in detail, since it has no adverse
biological activity, but it is now included to show the profiles that are obtained from the carbohydrates. Formation
profiles for H_2O from the thermal degradation of starch in
He and He-H_2O are presented in Figure 14. The maximum at
150°C most likely is due to the loss of water of hydration
(Fig. 5). The major quantity of water is formed during the
thermal degradation of starch via chemical dehydration
between 300° and 400°C. Above 400°C, water is formed continuously, but at greatly reduced level. It should be noted
that these data agree quite well with the formation profile
for water obtained by Shafizadeh [40].

The influence of water-saturated helium shows that the
temperature for maximum formation is approximately 25°C
higher than that obtained in dry helium. This is expected,
since the presence of water in the carrier gas will retard
the decomposition reactions that favor dehydration [45]. It
appears that the level of H_2O formed in water-saturated
helium is larger than in dry helium, but this is due primarily to the background water in the carrier gas. Adjustment
for the background water would show that level of water is

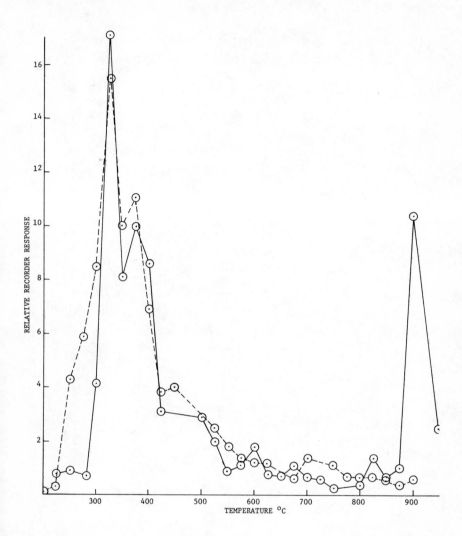

Fig. 13.   Formation profiles for carbon dioxide from dextrose; He (———), He-H$_2$O (- - -).

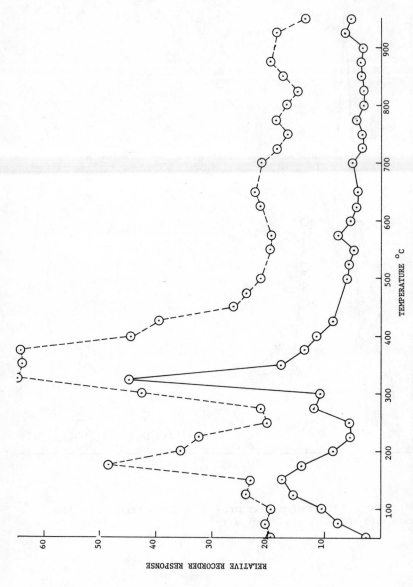

Fig. 14. Formation profiles for H_2O from starch; He(——), He-H_2O (- - -).

approximately the same as that obtained in a helium
atmosphere.

The influence of water on the formation of the low
molecular weight hydrocarbons does not show any general
trends, since only a small number of carbohydrates were
studied. Profiles for methane from cellulose (Fig. 15) show
that water has little influence on methane formation. On the
other hand, water can be effective in reducing the level of
the low molecular weight aldehydes and ketones. Figure 16 is
typical of the profiles obtained from the majority of the
carbohydrates. This profile clearly shows that water
decreases levels of acetaldehyde formed from cellulose, but
the level of reduction is low.

The data for the influence of water on the formation
of pyrolysis products are far from complete at the present
time. This portion of the study has shown that water influences the formation of carbon monoxide, the hydrocarbons, and
the low molecular weight aldehydes and ketones. To more
clearly establish the influence of water, a more complete
analysis of pyrolysis products must be obtained.

III. CONCLUSIONS

The results from this study on the thermal decomposition of carbohydrates and the analysis of several of the low
molecular weight products have added to the knowledge of the
thermal degradation of carbohydrates and plant materials.
The optimum formation for several gas phase constituents have
been determined, yet information is needed for additional
products in order to obtain a better understanding of the
formation of the thermal degradation products. The studies
on the thermal degradation of carbohydrates and the factors
that control the formation of individual pyrolysis products
using thermal analysis techniques are just beginning. Future
studies will answer many of the questions that have plagued
the chemists involved with pyrolysis studies on plant
materials.

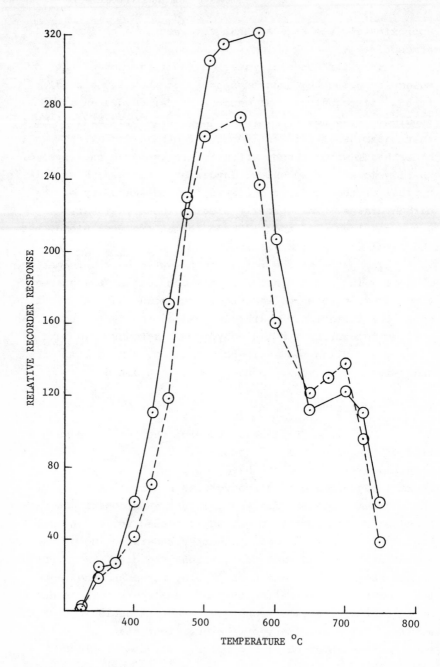

Fig. 15.   Formation profiles for methane from cellulose; He(——); He-H_2O (- - -).

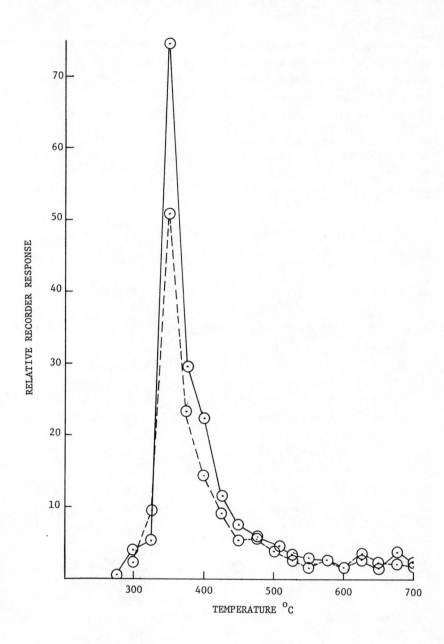

Fig. 16. Formation profiles for acetaldehyde from cellulose; He(——), He-H$_2$O (- - -).

ACKNOWLEDGEMENTS

The author thanks George Childs for assisting with this study, since much of this study would not have been possible without his valuable contributions. This study was supported (in part) by University of Kentucky Tobacco and Health Research Institute Project No. KTRB 23069.

REFERENCES

1. W. K. Tang and W. K. Neill, J. Polym. Sci. C 6, 65-81 (1964).
2. P. K. Chatterjee, J. Polym. Sci. 6, 3217-3233 (1968).
3. F. Shafizadeh, G. D. McGinnis, R. A. Susott, and H. W. Tatton, J. Org. Chem. 36, 2813-2818 (1971).
4. F. Shafizadeh, C. W. Philpot, and N. Ostojic, Carbohydrate Res. 16, 279-287 (1970).
5. K. Heyns and M. Klier, Carbohydrate Res. 6, 436-448 (1968).
6. S. Glassner and A. R. Pierce, III, Anal. Chem. 37, 525-527 (1965).
7. D. Gardiner, J. Chem. Soc. C 1966, 1473-1477 (1966).
8. I. S. Fagerson, J. Agr. Food Chem 17, 747-750 (1969).
9. H. W. Eickner, Forest Prod. J. 1962, 194-199 (1962).
10. F. L. Brown and W. K. Tang, Fire Res. Abstr Rev. 4, 76-91 (1962).
11. F. Shafizadeh and G. D. McGinnis, Carbohydrate Res. 16, 273-277 (1970).
12. H. R. Burton and D. Burdick, Tobacco Sci. 11, 180-185 (1967).
13. R. J. Phillipe, H. Moore, and V. Mazzone, Tobacco Sci 7, 21a-27 (1963).
14. M. D. Edmonds, M. T. Core, A. Bavley, and R. F. Schwenker, Jr., Tobacco Sci. 9, 48-53 (1965).
15. D. Burdick and H. R. Burton, Tobacco Sci. 12, 16-18 (1968).
16. H. R. Burton and S. Sheen, Unpublished results.
17. T. C. Tso, Preventive Med. 3, (1974).

18. E. L. Wynder and D. Hoffmann, "Tobacco and Tobacco Smoke," Academic Press, New York (1967).

19. R. W. Jenkins, Jr., R. H. Newman, R. D. Carpenter, and T. S. Osdene, Beitr. Tabakforsch. 5, 295-298 (1970)

20. R. R. Baker, Nature 247, 405-406 (1974).

21. G. P. Touey and R. C. Mumpower, Tobacco Sci. 1, 33-37 (1957).

22. R. R. Baker, High Temp. Sci. 7, 236-247 (1975).

23. R. R. Baker, and K. D. Kilburn, Beitr. Tabakforsch. 7, 79-87 (1973).

24. K. Kato, N. Takahashi, and Y. Kaburaki, Jap. Monop. Corp. Cent. Res. Inst. Sci. Papers 107, 165-169 (1965).

25. P. D. Garn, "Thermoanalytical Methods of Investigation," Academic Press, New York (1965).

26. W. W. Wendlandt, "Thermal Methods of Analysis," Wiley (Interscience), New York (1964).

27. P. D. Garn, Talanta, 11, 1417-1432 (1964).

28. J. A. Murphy, J. A. Hill, G. P. Schacher, Anal. Chem. 32, 1374-1375 (1960).

29. H. R. Burton, Beitr. Tabakforsch. 8, 78-83 (1975).

30. R. R. Baker, Beitr. Tabakforsch. 8, 16-27 (1975).

31. H. R. Burton and G. Childs, Jr., Beitr. Tabakforsch. 8, 174-180 (1975).

32. J. H. Bell, A. O. Saunders, and A. W. Spears, Tobacco Sci. 10, 305-313 (1961).

33. F. Shafizadeh, Advan. Carbohyd. Chem. 23, 419-474 (1968).

34. F. Shafizadeh, J. Org. Chem. 37, 278-284 (1972).

35. Y. Houminer and S. Patai, Tetrahedron Lett. 1297-1300 (1967).

36. K. Heyns, R. Stute, and H. Paulsen, Carbohydrate Res. 2, 132-149 (1966).

37. F. Shafizadeh and Y. Z. Lai, Carbohydrate Res. 42, 39-53 (1975).

38. F. Shafizadeh and Y. Z. Lai, Carbohydrate Res. 40, (1975).

39. R. R. Baker, J. Therm. Anal. 8, 163-173 (1975).

40. F. Shafizadeh, J. Appl. Polym. Sci. 28, 153-174 (1975).

41. G. A. Byrne, D. Gardiner, and F. H. Holmes, J. Appl. Chem. 16, 81-88 (1965).

42.   R. L. Stedman, <u>Chem. Rev.</u> <u>68</u>, 153-207 (1968).

43.   H. R. Burton and G. Childs, Jr., <u>Beitr. Tabakforsch.</u>, in press.

44.   H. R. Burton.   Unpublished results.

45.   F. Paulik, J. Paulik, and L. Erdey, <u>Acta. Chem. Acad. Sci. Hung.</u>, <u>26</u>, 143-148 (1961).

CONCLUDING DISCUSSION OF SALIENT ISSUES

K. V. Sarkanen

College of Forest Resources
University of Washington
Seattle, Washington

D. A. Tillman

Materials Associates, Inc.
Washington, D.C.

The papers presented at this symposium addressed a
wide variety of issues, including the production of energy
and chemicals from renewable resources. To that end, many
contributors presented additional useful, fundamental information. Others examined products that might become available from new and improved processes applied to biomass.
From these papers and the ensuing discussion, several key
issues emerged.

Energy from renewable resources is, perhaps, the most
significant problem of immediate concern; the production of
chemicals is a more long-term proposition. Within the energy
area, this country frequently forgets that biomass already
contributes some 1.3 quadrillion BTUs to the economy, with
forest industry residues supplying some 1.1 quads and the
remaining amounts coming from municipal wastes, bagasse, and

other waste products. This is a deliberate contribution. As
an example, the pulp and paper industry has invested millions
of dollars for the recovery of energy from Kraft black
liquor. Paper producers utilize these residues plus bark and
sawdust in the production of process steam and electricity.
In the process, the paper industry achieved an admirable
position; it produces 42% of its energy requirements.
Because this energy is not bought or sold as such, the U.S.
Government has failed to recognize its important contribution. It is a blind spot. Biomass, however, makes an energy
contribution that is comparable to that made by nuclear power.

Biomass, or renewable resources, can make an even
greater contribution in both the short and long run. In the
short run, forests can make more of a contribution. Presently, forest residues that are less than 4 inches in
diameter are not brought into yards for energy recovery.
They are not even counted as collectible residues. By changing these assumptions, forests could supply far more energy
than is currently anticipated. One example, provided by
R. Rothermel, is the lodgepole pine. This tree rarely grows
to beyond 4 inches in diameter. Its use could make a significant energy contribution. As another example of immediate
contribution, Dr. A. Broido pointed out that surveys of the
same tract of land using the same definitions varied by a
factor of 10 in the amount of waste generated. "When an
official proclamation comes out," says Broido, "nobody wants
to admit to inefficiency, so the low number is given."
Clearly, a more significant contribution to energy supply
could be made from silvicultural residues.

In the longer term, the collection of stubble from
harvested crop fields, improved forest stocking techniques,
and eventually, energy plantations could make significant
contributions. E. M. Wilson called for the development of
machines to ride behind the combine on a farm and harvest the
stubble for energy. Dr. D. Brink pointed out that in 25
years, forests could yield two to three times the fiber that
they do now--if improved stocking were to begin at the
present time. What is needed is engineering and design plus
improved planning and action to accomplish these contributions.

Conflicting with the need for improved planning and action is the nature of the United States. We are a nation of crisis solvers. We address energy problems when our supply is shut off. What we need is the immediate facing of the problem. If the United States examines the example of the Republic of South Africa, it finds a nation that decided to become energy independent. That nation took the necessary steps to make itself energy independent. This nation needs the guts to make such a decision and to take the steps to bring all sources of energy to bear--including renewable resources. Scientists and engineers must spread the word of the need to accomplish good planning at the government level to achieve this end.

Involved in the planning process and involved in the obtaining of action is the development of a realistic analysis. First, we must be realistic about the price of energy. This country has been spoiled by low-cost energy. Even the price of oil charged by the OPEC cartel is relatively low. We must get used to prices in the $3+/MM BTU range.

The planning process must also make a realistic assessment of biomass. The harvesting of renewable resources involves the use of large quantities of water. This harvesting also involves a large amount of transportation and therefore is energy intensive. Further, transportation energy is derived from petroleum. The problem is exacerbated. This realistic assessment must include a quantification of the effect on the soil of harvesting crop residues. The impact may be negligible if upwards of 80% of the residues are harvested, rather than plowed back for nutrients, but the quantification of this particular cost has not been performed. The effect is not known.

Even with the water, energy, and nutrient costs, biomass is not a net loser in energy production. The quantity of natural energy received is enormous, if one starts with the incidence of energy falling to the earth daily. Nature supplies most of the energy for biomass. Thus, even at 1 to 2% efficiencies, this approach to energy supply can be significant indeed. Further, when we consider that biomass now may carry a negative value, the process may appear very attractive. Brush in Southern California now grow and burn, causing extensive damage to property and homes. Harvesting this brush could eliminate the fire hazard and damage, which presently gives them a negative value. Similarly, some harvesting in forests may reduce conflagrations in the overaged forests. This recovery of biomass for energy production may convert negative value materials to positive value energy feedstocks. Clearly, the conversion of renewable resources can make a positive contribution to both energy and the environment if approached with realistic planning.

The energy contribution may come in various forms, including solid, gaseous, and liquid fuels. All such forms are being worked on, as the papers demonstrated. The selection of the best approach, however, must be accomplished on a case by case basis.

Dr. F. Shafizadeh points out that organic matter can be pyrolyzed into char, charcoal, and various chemical products. The solid fuels are high in energy content and low in sulfur. They would make an excellent fuel for power plants, and their transportation could be accomplished after conversion and avoid moving vast quantities of water and other impurities.

Gaseous fuels could be produced, for industrial or agricultural consumption, in several ways. These methods include the establishment of relatively small gasifiers within 35-100 miles of the source of organic residues. Gaseous fuels could be produced also by mixing the organic matter with coal and feeding it into a Lurgi gasifier designed for the production of substitute natural gas. This latter concept merits more research than is presently being performed.

Liquid fuels may also be produced in order to meet the needs of the transportation requirements. These fuels, however, are expensive to make. As. Dr. K. V. Sarkanen points out, energy conversion efficiencies rarely exceed 33%. Further, the fuels are lower in calorific content than petroleum. Finally, they can be corrosive, presenting materials handling difficulties.

Thus, the symposium concluded that renewable resources do make a substantial contribution to energy supply. It is a contribution that could be increased--perhaps to between 5 and 15 quads. It is a contribution which could approach 5% of the United States energy supply requirement in the year 2000. And, while this contribution could never make the United States energy self-sufficient, it could contribute significantly to an overall program designed to achieve that end.